Exercise Intolerance and Muscle Contracture

Springer

Paris
Berlin
Heidelberg
New York
Barcelona
Hong Kong
London
Milan
Singapore
Tokyo

G. Serratrice, J. Pouget, J.-Ph. Azulay (Eds)

Exercise Intolerance and Muscle Contracture

Springer

Georges Serratrice MD, FRCP
Professor of Neurology
Service de Neurologie et des Maladies Neuromusculaires
Centre Hospitalier Universitaire de la Timone
Marseille - France

Jean Pouget, MD
Professor of Neurology
Service de Neurologie et des Maladies Neuromusculaires
Centre Hospitalier Universitaire de la Timone
Marseille - France

Jean-Philippe Azulay, MD
Professor of Neurology
Service de Neurologie et des Maladies Neuromusculaires
Centre Hospitalier Universitaire de la Timone
Marseille - France

ISBN 2-287-59669-0 Springer-Verlag France, Berlin, Heidelberg, New York

Imprimé en France

Library of Congress Cataloging-in-Publication Data
Exercise Intolerance and muscle contracture / [edited by] G. Serratrice, J. Pouget, J.-P. Azuley
P. cm.
Includes bibliographical references and index.
ISBN 2-287-59669-0 (pbk.: alk. paper)
1. Contracture (Pathophysiology) 2. Exercise-Pathophysiology.
I. Serratrice, Georges. II. Pouget, J. (Jean), 1950-. III. Azulay, J.-P. (Jean-Philippe), 1960-.
[DNLM: 1. Contracture-physiopathology. 2. Exercise Tolerance-physiology. 3. Muscle, Skeletal-physiopathology. WE 545 E96 1999]
RC925.6.E94 1999
616. 7'407--dc21
DNLM/DLC
for Library of Congress 99-27078
CIP

SPIN: 10718590

Preface

The main themes presented in this volume are exercise intolerance and muscle contracture. These two topics could seem different at first sight but share common clinical features. For instance, symptoms of exercise intolerance are myalgias and cramps but metabolic contractures as well. Likewise exercise intolerance is sometimes due to dystrophinopathy, itself causing contractures. Therefore, it is justified to gather these two syndromes in one study.

Exercise intolerance and muscle contracture could seem well-known, even obsolete. It is not right. A reappraisal of these has become necessary for two reasons. First of all there are many new causes of exercise intolerance; secondly, the pathophysiology remains obscure in particular for muscle contracture and a clear classification is probably possible.

There is no book devoted to exercise intolerance and muscle contracture. This volume is divided in two parts related to the above topics. The two sections are introduced by a didactic and general overview, describing the main symptoms, classification and classical etiology. The other chapters, written by the best experts, present the recent advances in these syndromes and show many new aspects: new glycogenosis or mitochondriopathies, dystrophinopathies, malignant hyperthermia, myotonic disorders. Pathophysiology and classification of persistent contractures are described as well as therapy. Each chapter is thoroughly referenced, representing important progress in research in the field. Therefore, this comprehensive and updated volume allows neurologists, specialists in internal medicine, rheumatologists, pathologists, biochemists, physiotricians and all researchers interested in neuromuscular diseases to become familiar with many new and homogen data concerning exercise intolerance and muscle contracture which could really be considered as new syndromes. Moreover, this book will provide the basis for further developments. Most of the chapters have been presented in Marseille, France, on march 21st and 22nd 1998, during the XIV International Meeting on Neuromuscular Diseases.

G. Serratrice

Contributors

Andreu A.L.
H. Houston Merritt Clinical Research Center
Columbia University College of Physicians and Surgeons
630 West 168th Street
New York,
NY 10032, USA

Arahata K.
Department of Neuromuscular Research
National Institute of Neurosciences
National Center of Neurology and Psychiatry
Tokyo 187-8502, Japan

Aubert M.
Department of Anaesthesia and Intensive Care Unit
Hôpital d'Instruction des Armées
Laveran
13998 Marseille, France

Azulay J.-Ph.
Service de Neurologie et des Maladies Neuromusculaires
CHU de la Timone
13005 Marseille, France

Bartoli C.
Laboratoire de Biopathologie Nerveuse et Musculaire,
et Service d'Anatomie Pathologique et de Neuropathologie
Hôpital de la Timone,
13005 Marseille, France

Beckmann J.S.
Genethon,
Evry, France

Behan W.M.H.
University Department of Pathology
Western Infirmary
Glasgow G11 6NT, UK

Bendahan D.
Centre de Résonance Magnétique Biologique et Médicale
URA C.N.R.S. n° 6612
Faculté de Médecine, 27, Bd Jean Moulin
13005 Marseille, France

Bianco N.
Laboratoire de Biopathologie Nerveuse et Musculaire,
et Service d'Anatomie Pathologique et de Neuropathologie
Hôpital de la Timone
13005 Marseille, France

Bollini G.
Consultation des Maladies Neuromusculaires de l'Enfant
Service d'Orthopédie Pédiatrique
Hôpital d'Enfants, CHU Timone
13385 Marseille, France

Bruno C.
H. Houston Merritt Clinical Research Center
Columbia University College of Physicians and Surgeons
630 West 168th Street
New York, NY 10032, USA

Carrelet P.
Consultation des Maladies Neuromusculaires de l'Enfant
Service de Neurologie Pédiatrique
Hôpital d'Enfants, CHU Timone
13385 Marseille, France

Chabrol B.
Consultation des Maladies Neuromusculaires de l'Enfant
Service de Neurologie Pédiatrique
Hôpital d'Enfants, CHU Timone
13385 Marseille, France

Comi L.I.
Via del Pini, 101
80131 Napoli, Italy

Cozzone P.J.
Centre de Résonance Magnétique Biologique et Médicale
URA CNRS n° 6612
Faculté de Médecine, 27 Bd Jean Moulin
13005 Marseille, France

Deslangles O.
Laboratoire de Biopathologie Nerveuse et Musculaire,
et Service d'Anatomie Pathologique et de Neuropathologie
Hôpital de la Timone
13385 Marseille, France

Desnuelle C.
Service de Rééducation Fonctionnelle,
Hôpital de l'Archet
06200 Nice, France

Di Mauro S.
H. Houston Merritt Clinical Research Center,
Columbia University College of Physicians and Surgeons
630 West 168th Street
New York, NY 10032, USA

Dubas F.
Service de Neurologie A
Hôpital Larrey
CHU Angers
49033 Angers, France

Fardeau M.
I N S E R M 153, Institut de Myologie
Groupe Hospitalier La Salpêtrière
47, Bd de l'Hôpital
75651 Paris Cédex 13

Figarella-Branger D.
Laboratoire de Biopathologie Nerveuse et Musculaire,
et Service d'Anatomie Pathologique et de Neuropathologie
Hôpital de la Timone
13005 Marseille, France

Fujimoto S.
National Institute of Neurosciences
Tokyo 160, Japan

Gastaut J.-L.
Service de Neurologie
Hôpital Sainte-Marguerite
13009 Marseille, France

Griggs R.C.
University of Rochester
School of Medicine and Dentistry
Rochester, NY 14642, USA

Hadjigeorgiou G.M.
H. Houston Merritt Clinical Research Center
Columbia University College of Physicians and Surgeons
630 West 168th Street
New York,
NY 10032, USA

Haller R.G.
Neuromuscular Center
Institute for Exercise and Environmental Medicine of Presbytarian
Dallas Hospital
V A Medical Center and The University of Texas
Southwestern Medical Center
7232 Greenville Avenue
Dallas, TW 75231, USA

Ishikawa T.
Department of Pediatrics
Nagoya City
University Medical School
Nagoya 466, Japan

Janzer R.C.
Service de Neurologie et Institut de Pathologie
CHU Vaudois,
1101 Lausanne, Switzerland

Kakulas B.A.
Department of Neuropathology,
Royal Perth Hospital and the Australian
Neuromuscular Research Institute
University of Western Australia
Perth, Western Australia

Karpati G.
Montreal Neurological Institute and Hospital
3801 University Street
Montreal, Quebec
Canada H3A 2B4

Kozak-Ribbens G.
CRMBM-CNRS UMR 6612
Faculté de Médecine
27, Bd Jean Moulin
13005 Marseille, France

Kubo S.
Department of Neuromuscular Research
National Institute of Neurosciences
National Center of Neurology and Psychiatry
Tokyo 160, Japan

Kuntzer T.
Service de Neurologie et Institut de Pathologie
CHU Vaudois
1101 Lausanne, Switzerland

Layzer R.B.
University of California
San Francisco CA 94143 0114, USA

MacLennan D.H.
Montreal Neurological Institute and Hospital
3801 University Street
Montreal, Quebec
Canada H3A 2B4

Meola G.
Department of Neurology, University of Milan
San Donato Hospital
San Donato Milanese, Milan, Italy

Montfort M.F.
Laboratoire de Biopathologie Nerveuse et Musculaire, et Service d'Anatomie Pathologique et de Neuropathologie
Hôpital de la Timone
13005 Marseille, France

Moxley III, R.T.
University of Rochester
School of Medicine and Dentistry
Rochester, NY 14642, USA

Munsat T.L.
Department of Neurology
Tufts University
Boston, MA 02111, USA

Nigro G.
Via del Pini, 101
80131 Napoli, Italy

Nonaka I.
National Center Hospital for Mental
Nervous and Muscular Disorders
Tokyo 187-8551, Japan

Paquis V.
Laboratoire de Génétique
Hôpital de l'Archet
06200 Nice, France

Pellissier J.F.
Laboratoire de Biopathologie Nerveuse et Musculaire,
et Service d'Anatomie Pathologique et de Neuropathologie
Hôpital de la Timone
13005 Marseille, France

Pénisson-Besnier I.
Service de Neurologie A
Hôpital Larrey
CHU Angers
49033 Angers, France

Pouget J.
Service de Neurologie et des Maladies Neuromusculaires
CHU Timone
13005 Marseille, France

Richard I.
Genethon
Evry, France

Rowland L.P.
Columbia Presbyterian, Medical Center
The Neurological Institute
710 West 168th Street
New York, NY 10032-2603, USA

Saito K.
National Institut of Neurosciences
Tokyo 187-8502, Japan

Sansone V.
Department of Neurology, University of Milan
San Donato Hospital
San Donato Milanese, Milan, Italy

Serratrice G.
Service de Neurologie et des Maladies Neuromusculaires
CHU de la Timone
13005 Marseille, France

Shanske S.H.
Houston Merritt Clinical Research Center,
Columbia University College of Physicians and Surgeons
630 West 168th Street
New York, NY 10032, USA

Trojaborg W.
The Neurological Institute
Columbia Presbyterian, Medical Center
710 West 168th Street
New York, NY 10032-2603, USA

Tsuchiya Y.
Department of Neuromuscular Research
National Institute of Neuroscience
National Center of Neurology and Psychiatry
Tokyo 187-8502, Japan

Uyama E.
Department of Neurology
Kumamoto University School of Medicine
Kumamoto 860, Japan

Contents

Exercise Intolerance

Exercise Intolerance: Classification and Semiology

G. Serratrice, J. Pouget, J.-Ph. Azulay

Exercise intolerance syndrome includes several symptoms or signs: myalgias, cramps, weakness or myoglobinuria following exercise and relieved by rest. It results from various abnormalities of muscle metabolism.

Clinical symptoms

The cardinal feature is that normal exercise is impossible for the patient because of the development of disabling symptoms. These symptoms are of varying severity. The choice of the laboratory examinations depends on the clinical features more or less specific of the metabolic abnormality.

Physiological exercise intolerance

The quantification of exercise intolerance is difficult. It is possible for a normal subject to exercise to such an extent that the energy stores of muscle are markedly depleted. Fatigue is then prominent but muscle pain is usually absent. The ensuit fatigue protects the muscle against further exercise which might be harmful. Most of the patients consulting for exercise fatigue, muscle pain or cramps after strong exercise, have no metabolic deficiency. Progressively limited performances may be due to the lack of training or to physiologic changes linked with normal aging.

Myalgias and cramps (painful contractures)

Exercise provokes disabling symptoms and signs: myalgias, cramps when the energy supply is unable to sustain the exercise in a normal fashion due to the lack of an enzyme activity, glycolytic or lipidic, or to a mitochondrial metabolism abnormality. Such symptoms are defined as exercise intolerance.

The intensity of exercise varies from patient to patient: either brief and strong, or prolonged. The main symptoms are:

- Isolated myalgias with muscle pain more or less localized to exercising muscles otherwise generalized. Fatigue is associated with myalgia.

Sometimes exercise is possible after a short break, so-called the "second wind" phenomenon: the exercise tolerance increases tremendously if the patient is slowing down immediately after the initial sensation of fatigue.

- In more advanced cases pain is associated with cramps that are described as painful, unvoluntary, paroxysmal muscle contraction with stiffness of one or several muscles predominantly affecting the lower limbs (or exercising muscles), lasting for several hours, even overnight. In fact the word cramp may lead to confusion. Ordinary cramps are principally neurogenic with electromyographic bursts of high frequency. On the opposite, metabolic cramps, otherwise called "metabolic contractures" are also painful and brief, but the muscle contraction is electrically silent.

Fatigability, exercise weakness

Myalgias and cramps are in some patients associated with muscle fatigue or even with weakness and "intermittent muscle limping".

In such cases the intolerance is more severe, with repeated bouts of muscle pain, proximal weakness occurs during walking, followed by falls or lower limbs stiffness. The activity is severely curtailed and walking is arduous.

Exertional myoglobinuria

Myoglobinuria is the most severe form of exercise intolerance due to an extended rhabdomyolysis.

Urine is dark, usually brownish rather than red, when myoglobin is present in a sufficient amount (> 250 mg/ml). All the constituents of the necrotic muscle are discharged into the blood raising plasma levels of sarcoplasmic enzymes, aminoacids (especially taurine), creatine, creatinine, potassium, phosphate, urate. Hyperphosphatemia may lead to secondary hypocalcemia during the first days. Calcium enters the muscle fibers. Renal failure occurs due to heme induced nephropathy with anuria. Later there may be hypercalcemia because of increased secretion of parathormone and vitamin D, with deposition of calcium in injured muscle followed by release. Administration of mannitol or osmotic diuretics, alkalinisation of urine, volume replacement, dialysis, plasmapheresis are important.

Compartmental syndrome

Compartmental syndrome denotes the occurrence of symptom that results from the mechanical restriction of muscle in anatomic compartments bounded by bones and fascias. Most commonly, lower limbs compartments and especially anterior tibial loge, are involved. An exertional rhabdomyolysis may cause acute muscle swelling to an extent that muscle blood supply is compromised causing muscles ischemia. On the other hand chronic compartment syndrome results from shifts of water from the vascular to the intramuscular space during heavy exercise. The increased muscle volume in a fixed volume results in an abnormal increase of intramuscular pressure. Muscle blood flow is reduced and an exertional muscle pain occurs.

Ancillary examinations

Serum creatine kinase activity is sometimes high at rest. Myoglobinuria and myoglobinemia refer to muscle necrosis.

Lactic acidosis and increased lactate/pyruvate ratio are suggestive of a mitochondrial oxydation deficiency. In case of a block on the glycogenolytic pathway, lactacidemia does not raise during a forearm exercise under ischemia. Ammoniemia variations are less specific.

Exercise test

Exercice on an ergometric bicycle may contribute to differentiate exercise intolerance due to mitochondrial from those due to glycogenolytic defects. In case of oxidative defect, the incremental exercise test induces a decrease of VO_2 max (or maximum oxygen uptake), an earlier anaerobic threshold, an abnormal high level of lactate in serum during and after exercise, and a decrease of the work load. Both ventilation volume/oxygen consumption and ventilation volume/Co^2 production ratios increase.

Phosphorus Nuclear Magnetic resonance (31 P-NMR) as well as spectroscopy of muscle may also contribute to differentiate the various metabolic myopathies [1]. For instance, in glycolytic pathway block it shows lack of acidosis during exercise and in oxidative deficiency, a decrease of the ratio phosphocreatin/inorganic phosphorus, an increase of lactic acidosis and a slow recovery of pH and phosphocreatine after exercise.

Muscle biopsy is usually the most informative tool [2] by showing as main features: glycogen accumulation in muscle fibres in case of glycolytic enzymes deficiency, ragged red fibres, giant mitochondria and respiratory chain complex defect in mitochondrial myopathy.

Finally, a molecular genetic study can be proposed in familial and hereditary metabolic abnormalities.

Diagnosis

The diagnosis of exercise intolerance is often difficult. False exercise intolerance has to be excluded. Myasthenic fatigability is usually easily recognized. Cataplexia occurs more often spontaneously than during exercise. Gait disturbances, especially in elderly are worsened by exercise, particularly brachybasia.

Non muscular causes are lower limbs arteriopathy, spinal paraparesia. However, lumbar vertebral stenosis is often overdiagnosed. Certainly it does exist but only in cases with transient cauda equina syndrome, isolated myalgias or cramps are not related to stenosis of spine.

Rare patients may experience distal pain occuring in exercise associated with burning sensations hypohydrosis. This semiology is suggestive of Fabry's disease due to X linked alphagalactosidase A deficiency. Fabry's disease is probably often misdiagnosed in young men who complains of distal pains occuring after a mild exercise, of burning pain exacerbated by heating, alcohol and reduced by cold water. Anhydrosis is often present with lack of sudation. Angiokeratomas, corneal deposits are characteristics.

Muscular causes of exercise intolerance

The main causes are glycolytic and mitochondrial enzymes deficiencies [3, 4].

Glycolytic enzymes deficiencies

The most frequent are phosphorylase and phosphofructokinase deficiencies [5].

The breakdown of muscle glycogen is complex. Two enzymes are important in exercise intolerance syndrome: phosphorylase and phosphofructokinase. The phosphorylase is associated with a complicated activating system. It cleaves the alpha 1-4 links to start the process leading to the formation of pyruvate. Phosphorylase β is activated by phosphorylase β kinase. Phosphorylase α is then able to cleave glucose 1 phosphate. Glucose 1 phosphate produces glucose 6 phosphate by phosphoglucomutase; glucose 6 phosphate produces fructose 6 phosphate by phosphohexo-isomerase; fructose 6 phosphate produces fructose 1-6 diphosphate by phosphofructokinase. Phosphofructokinase is the controlling enzyme in the pathway to pyruvate. Its action is not reversible and synergic to phosphorylase. Phosphorylase "push" glycogen breakdown whereas phosphofructokinase "pull out" it and controls the flow through this metabolic sequence.

Phosphorylase deficiency

Phosphorylase deficiency corresponds to glycogenosis type V (Mc Ardle's disease). The illness is inherited as an autosomal recessive trait or rarely a dominant. The locus is on chromosome 11 (11q 13). The gene product is phosphorylase.

Usually the onset is in the childhood. Before 10 year-old, patients complain of fatigue and inability to keep up with others of his age. Aching in the legs is noted in early adolescence. An heavy bout of exercise is accompanied by painful cramps (metabolic contractures) in the muscle. The pain will last for several hours, even overnight. The urine become dark because of myoglobinuria. With the beginning of exercise a rapid acceleration of heart rate and of breathing occur.

The "second wind" phenomenon is usual and after a sense of laboring and tightness in the muscles, the patient has a sense of ease and relaxation and is able to go on exercise for a longer period.

With time, permanent proximal muscles weakness and atrophy may occur, especially in shoulder girdle. Myoglobinemia is elevated in 50% of cases.

There is an absence of rise of lactate levels in the venous blood from the exercised forearm under ischemic condition. No lactate is produced. Serum creatine kinase is normal or mildly increased at rest. EMG in the tightened muscles reveals the absence of electrical activity. It is a true contracture. In case of a late, permanent weakness, EMG may be myogenic.

Muscle biopsy shows subsarcolemmal or sometimes intermyofibrillar deposits of PAS stained glycogen. Vacuolar myopathy aspects are rare. Scattered necrotic fibres are common. Biochemical measurement of glycogen shows an increase from the normal 1% to around 2 or 3% of muscle weight. Histochemical study of phosphorylase is negative in muscle fibres, positive in vessel walls.

Resonance magnetic spectroscopy [6] reveals a normal resting phosphorus magnetic resonance. There is no decrease in intramuscular pH in response to ischemic exercise. Phosphocreatine/inorganic phosphorus ratio is increased with normal recovery.

Contracture was initially thought to be similar to rigor mortis. The exercise would deplete ATP because the muscle fiber could not generate ATP through the glycolytic pathway under anaerobic conditions. Without ATP the cross-bridge of the contractile-protein would stop and lock in a rigor state. This explanation was not true and several studies showed that ATP was not depleted during exercise in patients with Mc Ardle's disease [6, 7]. Experimental blockade of glycolytic enzymes [8] showed ADP increase, increased concentration of intracellular calcium which, with a lack of acidification, could increase the calcium sensitivity of the contractile protein.

The concentration of muscle Na^+ K^+ pumps (studied by oubaïn binding) is low in Mc Ardle's disease compared with controls [9]. This low pump concentration promotes extracellular potassium accumulation. It contributes to a development of fatigue attributable to membrane inexcitability at high rate of muscle concentration. The increase in potassium release promotes exaggerated potassium mediated autonomic responses especially cardiovascular, exercise heart rate in particular. The mechanism of the low Na^+ K^+ pump concentration is unknown. In healthy subjects, increase in pumps quantity has been linked to physical training and lower quantity of pumps to physical inactivity. A relative physical inactivity could be the physiopathologic bases in Mc Ardle's disease but this hypothesis seems unprobable [9]. The high level of plasma potassium could promote an increase in pumps, if they are normal. Despite this stimulus, no modification is produced.

The nature of the "second wind" phenomenon is due to a change in the supply of fatty acid to the muscle. It could be associated with change in the blood flow to the muscle.

Phosphofructokinase deficiency

Clinically, this illness closely resembles Mc Ardle's disease for the obvious reason that the defect is in the same chain of metabolic reaction.

The pattern of inheritance is autosomal recessive. The gene is located on chromosome 1 (1q 32); the gene product is phosphofructokinase, and the disease is quoted as type VII glycogenosis or Tarui's disease.

The patient, more often a man, experiences the early onset of myalgias, lower limbs metabolic contracture associated with fatigue and often nausea or vomiting. Sometimes fatigue predominates on upper limbs giving difficulties in playing piano, for example, in some patients. With strenuous exertion, severe muscle pains, contractures and myoglobinuria occur [10]. The "second wind" phenomenon is rarely experienced. Serum creatine kinase (CK) level is high after exercise or even between exercises at rest. There is no rise of venous lactate following ischemic forearm exercise and muscle cramps may appear. Contractures are silent, EMG may sometimes show myogenic abnormalities with fibrillations. Phosphofructokinase is absent in muscle, decreased in erythrocytes. Muscle biopsy shows PAS + subsarcolemmal glycogen accumulation as well as abnormal polysaccharide deposits, perhaps an insoluble form of glycogen. Biochemical dosages show elevated concentrations of glucose 1 phosphate and glucose 6 phosphate; concentration of fructose 1 6 diphosphate is reduced.

Magnetic resonance spectroscopy shows a decrease of ATP at rest (with high ratio phosphocreatine/ATP) and during exercise [1]; pH is normal. After exercise there is an abnormal accumulation of monophosphate esters.

Other glycolytic deficiencies

They are unfrequent. However, some of them may induce exercise intolerance. [11]

Phosphoglycerate kinase deficiency [12] The inheritance is X linked recessive (Xq 13) in childhood or adolescence. Pains and lower limbs weakness occur during exercise associated with myoglobinuria. Serum creatine kinase level is highly elevated. Forearm ischemic exercise test produces no rise in venous lactate.

Phosphoglycerate mutase deficiency [13] Sometimes or in some patients myalgias, metabolic contractures and recurrent myoglobinuria may follow exercise. The gene abnormality is located on chromosome 7 (7p12 p13).
Lactate deshydrogenase deficiency Myalgias, cramps and myoglobinuria are observed at exercise. Serum CK is elevated. The gene abnormality is located on chromosome 11 (11p 15-4).

Aldolase deficiency is a cause of exercise intolerance, weakness, serum creatine kinase elevation associated with intercurrent fever [14]. Glucose 6 phosphate dehydrogenase deficiency which causes acute hemolytic crises after ingestion of fava, is another cause of exercise intolerance [15] by the mechanism of pentose phosphate pathway defect.

Mitochondrial enzymes deficiencies

Mitochondrial diseases are clinically heterogeneous and difficult to classify because of the many overlapping symptoms and signs. Nevertheless, exercise is a classical manifestation of mitochondrial myopathy and may be observed either isolated or combined with other symptoms.

The main function of mitochondria is to produce ATP, which is the universal energy purveyor. Mitochondrial metabolism is divided in five steps:
- transmembrane transport
- oxidation
- Krebs' cycle
- oxidation/phosphorylation coupling
- respiratory chain.

Depending on the intensity of exercise (prolonged exercise, brief intense effort), the fuel for mitochondrial energy production shifts between carbohydrates and fatty acids. Mitochondrial carbohydrate oxidation starts with cytoplasmic pyruvate. Pyruvate is transported across the inner mitochondrial membrane through the action of a translocase and is then oxidatively decarboxylated by the pyruvate dehydrogenase complex (PDHC), which produces Co_2 and acetylCoA, while reducing one NAD to NADH, H^+. Acyl CoA is further oxidized through the Krebs' cycle leading to the production of NADH, H^+ and FADH, H^+. Fatty acids are stored in the cytoplasm as triglycerides in lipid droplets which are often adjacent to mitochondria. When required, they are released by a triglyceridelipase and energized to acylCoA by an acylCoA lipase on the mitochondrial outer membrane. The translocation of long-chain acylCoA across the inner mitochondrial membrane requires carnitine, carnitine palmitoyltransferase (CPT) I and II, and a carnitine-acylcarnitine translocase. In the mitochondrial matrix, β oxidation reduces the length of the fatty acid chain by two carbons at each "pass", with liberation of acetylCoA, NADH, H^+ and FADH, H^+. NADH, H^+ and

FADH, H^+ derived from the oxidation of fat or sugars which enter the mitochondrial electron transport chain; hydrogens from these compounds go through a serie of oxidation-reduction reactions, eventually reducing molecular oxygen to water. Energy released during these reactions is utilized to pump hydrogen ions through the inner mitochondrial membrane which creates a transmembrane charge and pH differential.

The main sites of respiratory chain are:

- Complex I (NADH coenzyme Q reductase) carries hydrogen from NADH to coenzyme Q. It contains 25 peptides.
- Complex II (succinate ubiquinone reductase) carries electron from NADH to coenzyme Q (which acts as a shuttle). It is composed of 5 protein subunits.
- Complex III (coenzyme Q cytochrome C reductase) oxides coenzyme Q. It is composed of 11 subunits.
- Complex IV (CoX, cytochrome oxidase) carries electrons from complex III to oxygen. It is composed of 13 proteins and 2 copper atoms.
- Complex V (ATP synthetase) allows some protons to flow back into the mitochondria and uses the energy so generated to synthetize ATP.

Exercise intolerance related to mitochondrial abnormalities differs from glycogenosis: fatigue is more frequent than cramps, myoglobinuria is rare. These cases can be divided in 4 groups according to the area of mitochondrial metabolism affected.

Defect of transport: carnitine palmiotyltransferase [16]

Inheritance is autosomal recessive. The gene abnormality is on chromosome 11 (11p 11p 13) [17].

The onset is early in childhood or in young adults. Recurrent episodes are triggered by prolonged exercise, fasting, insomnia and sometimes infectious disease [18]. Patients experience myalgias, weakness during exercise and myoglobinuria. Renal failure may occur and prolonged fasting may induce a serum CK level increase. Muscle biopsy is either normal or may show aspects of a lipidic myopathy. The defect involve CPT II. Glucidic diet is sometimes effective.

CPT II defect could be mimicked by some fatty acid betaoxidation enzymatic defect: acylcoA dehydrogenase of very long chain fatty acid [19], trifunctional protein which catalyzes terminal beta oxidation reactions [20, 21].

Defect of substrate utilization

Exercise intolerance is especially observed in childhood in conjunction with other features.

It is observed in pyruvate decarboxylase deficiency sensitive to thiamine. Exercise-induced myoglobinuria is present if a defect of betaoxidation of fatty acids is present (defect in acylcoA dehydrogenase long chain acids). Another disease is glutaric aciduria type II [22], which may cause a mild myopathy, starting in childhood or even in adult life, characterized by recurrent episodes of weakness and exercise intolerance accompanied by vomiting, hypoglycemia and acidosis.

Lipoamide dehydrogenase deficiency is a cause of episodes of muscle weakness and "anicteric hepatitis" in adult [23]. This enzyme is a catalytic protein of the pyruvate dehydrogenase complex which convert pyruvate to acylco A.

Defect of Krebs' cycle

Aconitase deficiency is associated with exercise intolerance and myoglobinuria [24]. A complex II deficiency is present.

Defects of the respiratory chain

Bouts of exercise myalgias start in childhood or adult life in complex I deficiency. Myalgias, cephalalgias, vomitings occur climbing stairs. These cases improved by riboflavin but may be later complicated by proximal weakness. The lactacidemia is elevated. Muscle biopsy shows ragged red fibers and abnormal mitochondrias as well as respiratory chain enzymes deficiency. Magnetic resonance spectroscopy is abnormal. At rest the ratio phosphocreatine/phosphorus is low and during exercise the slope decreases. After exercise phosphocreatine recuperation is slow.

Childhood exercise myalgias with hyperlactacidemia are observed in complex II and III deficiency. In case of coenzyme Q_{10} deficiency [25, 26] lipidic myopathy, exercise intolerance, epilepsy and cerebellar syndrome have been described. In complex III, deficiency myopathy with weakness and exercise intolerance is accompanied by cytochrome b deficiency. C and K vitaminotherapy could be effective. In complex IV a cytochrome oxidase subunit III defect may cause recurrent myoglobinuria [27].

Abnormalities of calcium in the sarcoplasmic reticulum

During muscle contraction, electric membrane stimulation produces a release of the ion calcium sequestred in sarcoplasmic reticulum cisternal, through a calcium release channel, the ryanodine receptor. During the relaxation of muscle, a reuptake of calcium ion toward sarcoplasmic reticulum is assumed by calcium ATPase. Malignant hyperthermia and Brody's syndrome both arise from primary abnormalities of calcium binding by the sarcoplasmic reticulum. Malignant hyperthermia (related in 50% of familial forms to ryanodine receptor gene mutation on chromosome 19) is attributed to a prolonged open time of the calcium releasing channel. Brody's syndrome is attributed to an altered ATPase of the sarcoplasmic reticulum.

Excessive release: exercise intolerance and malignant hyperthermia

Malignant hyperthermia, previously familial disease alerting anesthesiologists, recently became also a neurological disease, observed in sporadic cases of exercise intolerance without known etiology [28]. Malignant hyperthermia is a model of excitation/contraction coupling deficiency, due to excess of calcium release by sarcoplasmic reticulum.

Normally the sarcoplasmic reticulum binds calcium (by a calcium ATPase) during relaxation. It releases the calcium through a channel, the ryanodine receptor (different from calcium ATPase), and channel opening time is brief. Calcium induces contraction and glycolysis.

In familial autosomal dominant malignant hyperthermia ryanodine receptor gene mutation, located on chromosome 19, is found in 50% of cases. This mutation induces a structural defect of ryanodyne receptor. Its open time become prolonged, and leading to calcium accumulation. This excess of calcium by several steps is resulting in excessive heating, acidosis, muscle rigidity itself producing hyperthermia. High level of calcium stimulates phosphorylase kinase, increasing glycolysis and lactic acid production. Due to hyperthermia, sensitivity to calcium of contractile proteins increases. Calcium induces contraction and contracture. Loss of ATP does not allow actin-myosin dissociation, reduces calcium ATPase activity. Oxidative phosphorylation is reduced and heat production increases. There is a leak of creatine kinase, potassium and myoglobin.

This phenomenon occurs in sporadic rhabdomyolysis, different from familial malignant hyperthermia. These sporadic cases occur with or without hyperthermia.

Exertion syndrome with hyperthermia

This syndrome occurs after sudden or heavy exercise, frequently in military or body builders or untrained subjects [29]. A high heat, alcohol, a loss of salt are inducing factors, as well as hyposudation (mucoviscidosis, ichtyosis, ectodermal dysplasia). Other factors are status epilepticus, status asthmaticus, strychnin intoxication, muscle glycogenosis, thalassemia, mitochondriopathies.

The onset of attack is sudden; at exercise temperature is high. Myalgias, muscle pains, fatigue and cramps occur; then a severe exercise intolerance syndrome [30] is accompanied by vomiting, muscle weakness and stiffness, myoglobinuria, unability to walk and sometimes seizures. Neurological signs are prominent: transcient coma, consciouness abnormalities, seizures, cerebellar syndrome, hemiplegia but a cardiovascular collapsus, skin hemorrhages, hematomas, massive myoglobinuria with acute renal failure may also occur in severe cases. Serum creatine kinase is elevated; myoglobinemia, acidosis, hepatic enzymes are high with hypokaliema, hypocalcemia, hypophosphoremia. There is a sensitivity to halothane and cafein tests. Magnetic resonance spectroscopy [1, 31] shows phosphorus 31 abnormalities. Muscle biopsy shows rhabdomyolysis. Most often the course is favourable, consciousness abnormalities are transient. Hydratation and Dantrium are useful.

Forms without hyperthermia

In some young adults exercise intolerance without known metabolic cause, rarely with toxic cause [32] is related to malignant hyperthermia because of positive halothane and cafein tests. The symptoms occur either after brief and heavy exercise or after prolonged walk with fatigue, cramps, muscle stiffness. Creatine kinase is elevated at rest as well at exercise. Myoglobinuria is frequent. Amyotrophy may occur. Moderate doses of Dantrium are useful.

Undetermined cases of rhabdomyolysis with halothane and cafein tests positivity are described: post-viral, iatrogenic, toxic, recurrent.

Deficiency in reuptake

Brody [33] described in 1969 a rare syndrome with, as essential findings, a slowness of relaxation and impairment of calcium uptake by sarcoplasmic reticulum after exertion. Sarcoplasmic reticulum plays an important role in excitation/contraction coupling. In response to excitation of the T tubules, ion calcium sequestred in sarcoplasmic reticulum cisternal is released into the myofibrillar compartment. This calcium triggers molecular events that ultimately lead to myofilaments sliding and force generation. Subsequently the calcium content of the myofibrillar department is rapidly restored to the resting level by uptake of calcium into the sarcoplasmic reticulum, through the activity of calcium ATPase located in the sarcoplasmic reticulum membrane. The normal functioning of this pump is critical for the normal relaxation of muscle fibres. A reduction of sarcoplasmic reticulum calcium ATPase cause the Brody's syndrome manifestations.

Brody's syndrome is rare [34]. Inheritance is variably dominant autosomal, recessive or sporadic. There is a male predominance. The onset is during the 2nd or 3rd decade. Patients present with rapidly developing exercise stiffness and moderate weakness. Upon cessation of exercise, relaxation of distal and facial muscle is delayed. After contraction patients experience some difficulties in opening clenched first mimicking myotonia; however, electromyography does not show any myotonic discharges but only normal Motor Unit potentials. Cramps are frequent. Lactacidemia, normal at rest, increases at exercise. Muscle biopsy usually shows somes fibres necrosis, type II fibres atrophy and lack of immunoreactive sarcoplasmic reticulum calcium ATPase protein in type

II fibers. In addition Ca_2 uptake is reduced in isolated sarcoplasmic reticulum. *In vivo* phosphorus 31 shows excessive acidosis at exercise, fast decrease of phosphocreatine, normal recuperation of ADP and phosphocreatine. A mutation of ATP_2 A_1 gene (which encodes $SERCA_1$) is found on chromosome 16 (16q 12). This mutation produces a defect of fast sarcoplasmic reticulum ATPase. Calcium concentration increases in cytosol producing cramps, necrosis, stimulation of anaerobic glycolysis, contracture and pain. Therapy is based on the administration use of Dantrium and Verapamil.

Exercise intolerance and muscular dystrophy

Obviously all the muscle diseases may more or less produce some difficulties in exertion. Besides more and more cases of isolated exercise intolerance have been recently attributed to dystrophin abnormalities.

Dystrophinopathies

When a metabolic defect cannot be detected in a patient with exercise intolerance it is justified to study dystrophin. Some cases are asymptomatic and an isolated hyperCKemia is the only abnormality. These cases were previously considered as idiopathic hyperCKemia of unknown origin. In other patients, an exercise may precipitate acute muscle necrosis with myoglobinuria, elevated serum creatine kinase and in some cases massive renal failure and anuria. Finally, other patients complain of muscle stiffness, cramps, exercise myalgia.

The cases of myalgias, cramps, exercise intolerance without weakness associated with dystrophin deficiency are classified as type IV of dystrophinopathy and gene deletions were shown to be localized between exons 12 - 44 [35]. The other types of dystrophinopathies are type I (Duchenne muscular dystrophy), type II (Becker dystrophies), type III (limb girdle, quadriceps, distal forms with lack of dystrophin), type V (dilated cardiomyopathy [35]). However, some patients considered first as type IV cases may progress after several years of evolution to a chronic muscular dystrophy.

Exercise intolerance and myotonic syndromes

Myotonia is not a cause of exercise intolerance and is usually painless, in myotonic dystrophy or periodic paralysis. However, the molecular characterization of the myotonic disorders has identified new diseases, a number of which are characterized by muscle pain limiting activity and causing exercise intolerance. Sodium channelopathy (myotonia fluctuans, myotonia permans, periodic paralysis), and especially Proximal Myotonic Myopathy (PROMM) could induce exercise intolerance.

Undetermined significance abnormalities

Fishbein syndrome or myoadenylate deaminase deficiency

Described in 1978, this defect [36] may be present in several well defined neuromuscular diseases. However, a primary myoadenylate deficiency could be clinically characterized when an exercise intolerance with myalgia, cramps and sometimes myoglobinuria [37] are present, and biochemically caracterized by negligible residual activity in muscle [38].

Myoadenylate deaminase is one of the enzymes of the purine nucleotide cycle whose functions are: the stabilization of the energy potential, the activation of the glycolysis through the action of IMP on phosphorylase and of ammonia on phosphofructokinase, the activation of the buffer of the lactate's produce during exercise and the production of the Krebs' cycle intermediates such as fumarate which is formed in the anabolic phase of the purine nucleotide cycle. However, the functional significance of each of these functions during exercise *in vivo* has been questionned.

The cDNA and the gene have been cloned and sequenced; the gene has been localized to the short arm of chomosome 1 near the CPT gene.

Serum CK level at rest is usually normal but activity may variably increase the basal values. The forearm ischemic exercise produces a normal rise of venous lactate but no rise of ammonia and of inosine monophosphate, which are the products of myoadenylate desaminase reaction. EMG is usually normal or may show non-specific myopathic changes.

In some patients myoadenylate deaminase deficiency is the unique abnormality. These patients are adults, with a predominance of affected males. The disease is transmitted as a sporadic or an autosomal dominant trait [39]. The exercise intolerance is prominent manifested by myalgias, burning legs, cramps, fatigue, stiffness, and sometimes weakness induced by exercice. The course is chronic. Muscle biopsy is normal.

Some patients have combined genetic defects (most often associated with phosphorylase or phosphofructokinase deficiency). This "double trouble" seems to exacerbate the phenotypes [40].

Exercise intolerance with tubular aggregates

Tubular aggregates are non-specific morphological changes, observed in several neuromuscular diseases (periodic paralysis [41], myasthenia, congenital myopathy, intoxications, muscular dystrophies). They are unusually frequent in exercise intolerance syndrome [42, 43, 44, 45] in which they may sometimes constitute the major and the unique structural alteration. By light microscopy, they appear as a basophilic material with hematoxylin and eosin staining, localized in the subsarcolemmal and intermyofibrillar. They are stained by the modified Gomori trichrome technique, accentuated by NADH-TR and myoadenylate deaminase. Conversely they do not react with myofibrillar ATPase, SDH, or menadione linked alpha glycerophosphate dehydrogenase (GPDH). Tubular aggregates are exclusively localized in type II fibers in contiguity with sarcoplasmic reticulum. Electron microscopy shows that TA are localized under the plasma membrane or with an intermyofibrillar distribution. TA tubules are double walled with a second tubule of 50-70 nm in diameter and one or several inner tubules of 20-30nm in diameter.

Exercise-induced cramps and myalgias in which tubular aggregates are the only morphological abnormality are well-documented in sporadic cases. Patients are men over twenty complaining of pain and stiffness in the lower limbs [46] when walking or running. Myalgias are predominant in the lower limbs, sometimes in the upper limbs without any weakness or atrophy. EMG and serum CK level are normal. Muscle biopsy shows tubular aggregates confined to type II fibres.

Tubular aggregates correspond to extensions of the sarcoplasmic reticulum sacs as shown by electron microscopy. Immunohistochemical studies with a polyclonal antibody against Ca^{2+} SR ATPase show an immunoreactivity. Release of calcium from the sarcoplasmic reticulum during the muscle contraction is determined by Ca^{2+} transport ATPase, as well as during relaxation by active pumping of calcium ions through the sarcoplasmic reticulum. The question of calcium storage is controversial. Salviati et al. [47] incubated skinned muscle fibers in a calcium-oxalate medium and the ability of TA to store calcium was investigated by electron microscopy. They observed the formation of intraluminal Ca-oxalate deposits both in the terminal cisterns of the sarcoplasmic reticulum and in tubular aggregates. Different results obtained by others [42] in a quantitative analysis carried out on frozen sections stained from calcium by the Von Kossa method, showed that the calcium content density was lower than in normal sarcoplasm. These results obtained in physiologic conditions contrasted with the experimental conditions of the study of Salviati et al. [47] suggesting that tubular aggregates are not liable to an increased calcium uptake.

There is no treatment to exercise-induced cramps and myalgias. The course is benign [46]. Dantrolene sodium which allows the release of calcium from the sarcoplasmic reticulum is ineffective as well as calcium channel blockers.

References

1. Argov Z, Bank WJ (1991) Phosphorus Magnetic resonance spectroscopy (^{31}PMRS) in neuromuscular diseases. Ann Neurol 30:90-97
2. Karpati G, Arnold D, Matthew P, Carpenter S, Anderman F, Shoubridge E (1991) Correlative multidisciplinary approach to the study of mitochondrial encephalomyopathy. Rev Neurol 147:455-461

3. Di Mauro S, Tonin P, Servidei S (1992) Metabolic myopathies. Handbook of clinical neurology (vol 18). In: Rowland LP, Di Mauro S (eds) Myopathies. Elsevier, Amsterdam, pp 479-526
4. Di Mauro S (ed) (1992) Symposium on mitochondrial encephalomyopathies. Brain Pathol 2:153-205
5. Di Mauro S, Tsujino T (1994) Non lysosomal glycogenoses. In: Engel AG, Franzini-Armstrong C (eds) Myology. 2nd edn, McGraw-Hill, New York, vol 2, pp 1619-1628
6. Argov Z, Bank WJ, Maris J, Chance B (1987) Muscle energy metabolism in McArdle's syndrome by in vivo phosphorus magnetic resonance spectroscopy. Neurology 37:1720-1724
7. Rowland LP, Araki S, Crmel P (1965) Contracture in McArdle's disease: stability of adenosine triphosphate during contracture in phosphorylase-deficient muscle. Arch Neurol 13:541-544
8. Brumback RA (1980) Iodoacetate inhibition of glyceraldehyde-3-phosphate dehydrogenase as a model of human myophosphorylase deficiency (McArdle's disease) and phosphofructokinase deficiency (Tarui's disease). J Neurol Sci 8:383-398
9. Haller RG, Clausen T, Vissing J (1998) Reduced levels of skeletal muscle NA^+-K^+ ATPase in Mc Ardle's disease. Neurology 50:37-40
10. Tonin P, Levis P, Servideis S, Di Mauro S (1990) Metabolic causes of myoglobinurin. Am Neurol 27 : 181-185
11. Di Mauro S, Bresolin N (1994) Newly recognized defects of terminal glycolysis. In: Engel AG, Franzini-Armstrong C (eds) Myology. 2nd edn, McGraw-Hill, New York, vol 2, pp 1619-1628
12. Di Mauro S, Dalakas M, Miranda RF (1981) Phosphoglycerate kinase deficiency: a new cause of recurrent myoglobinuria. Ann Neurol 10:90
13. Di Mauro S, Miranda AF, Olarte M, Friedman R, Hays AP (1982) Muscle phosphoglycerate mutase deficiency. Neurology 32:584-591
14. Kreuder J, Borkhardt A, Repp R et al (1996) Inherited metabolic myopathy and hemolysis due to a mutation in aldolase. New Engl J Med 334:1100-1104
15. Bresolin N, Bet L, Moggio M et al (1987) Muscle G6PD deficiency. Lancet I: 212-213
16. Zierz S (1994) Carnitine palmitoyltransferase (CPT) deficiency: In: Engel AG, Franzini-Armstrong C (eds) Myology. 2nd edn, McGraw-Hill, New York, vol 2, pp 1619-1628
17. Minoletti F (1992) Localisation of the human gene for carnitine palmitoyltransferase to 1p13-p11 by non-radioactive in situ hybridization. Genomics 13:1372-1374
18. Desnuelle C, Pellissier JF, De Barsy Th, Serratrice G (1991) Intolérance à l'exercice par déficit en carnitine palmityl-transférase. Rev Neurol 146:231-234
19. Smelt AH, Poorthuis BJ, Onkenhout W et al (1998) Very long chain acyl-coenzyme A dehydrogenase deficiency with adult onset. Ann Neurol 43:540-544
20. Miyajima H, Orii KE, Shindo Y et al (1997) Mitochondrial trifuntional protein deficiency associated with recurrent myoglobinuria in adolescence. Neurology 49:833-837
21. Schaefer J, Jackson S, Dick DJ, Turnbull DM (1996) Trifunctional enzyme deficiency: Adult presentation of a usually fatal beta-oxidation defect. An Neurol 40:597-602
22. De Visser M, Scholte HR, Schutgens RBH, Bolhuis PA, Luyt-Houwen IEM, Vaandrager-Verduin MHM, Veder HA, Oey PL (1986) Riboflavin-responsive lipid-storage myopathy and glutaric aciduria type II of early adult onset. Neurology 36:367-372
23. Elpeleg ON, Saada AB, Shaag A et al (1997) Lipoamide dehydrogenase deficiency: A new cause for recurrent myoglobinuria. Muscle and Nerve 20:238-240
24. Hall RE, Henrikson KG, Lewis SF, Haller RG (1993) Kennaway NG: Mitochondrial myopathy with succinate dehydrogenase and aconitase deficiency. Abnormalities of several iron-sulfur proteins. J Clin Invest 92:2660-2666
25. Servidei S, Spinazzola A, Crociani P et al (1996) Replacement therapy is effective in familial mitochondrial encephalomyopathy with muscle CoQ10 deficiency. Neurology 46:A420

26. Keightley J, Hoffburn KC, Burton MD (1995) A microdeletion in cytochrome c oxidase subunit II associated with CoX deficiency and recurrent myoglobinuria. Nat Genet 12:410-416
27. Ogasahara S, Engel AG, Frens D, Mack D (1989) Muscle coenzyme Q deficiency in familial mitochondrial encephalomyopathy. Proc Nat Acad Sci Usa 86:2379-2382
28. Serratrice G, Kozak-Ribbens G, Cozzone PJ (1997) Aspects neurologiques de l'hyperthermie maligne. Rev Neurol 153:304-313
29. Aubert M, Petrognani R, Deslangles O (1993) Hyperthermie maligne d'effort. In: Aubert M, Borsarelli J, Kozak-Ribbens G (eds) Hyperthermies malignes. Normed Verlag, Bad Homburg Englewood Madrid, pp 5-14
30. Denborough MA, McLean A, Morgan G, Hopkinson KC (1994) Fatal inherited rhabdomyolysis and malignant hyperthermia. Lancet 343:236-237
31. Kozak-Reiss G, Gascard JP, Hervé Ph (1988) Malignant and exercise hyperthermia: investigation of 73 subjects by contracture tests and P31 NMR spectroscopy. Anesthesiology 69:A415
32. Anetseder M, Hartung E, Klepper S, Reichmann H (1994) Gasoline vapors induce severe rhabdomyolysis. Neurology 44:2393-2395
33. Brody I (1969) Muscle contracture induced by exercice. A syndrome attributable to decreased relaxing factor. N Eng J Med 281:187-192
34. Karpati G, Charuk J, Carpenter S, Jablecki C, Holland P (1986) Myopathy caused by deficiency of Ca^{2+} Adenosine Triphosphatase in Sarcoplasmic reticulum (Brody's disease). Ann Neurol 20:38-49
35. Samaha FJ, Quinlan JD (1996) Myalgias and cramps. Dystrophinopathy with wide ranging laboratory findings. J Child Neurol 11:21-24
36. Fishbein WN, Armbrustmacher VW, Griggin JL (1978) Myoadenylate deaminase deficiency a new disease of muscle. Science 200:545-548
37. Bruno C, Minetti C, Shanske S et al (1998) Combined defects of muscle phosphofructokinase and AMP deaminase in a child with myoglobinuria. Neurology 50:296-298
38. Serratrice G, Pellissier JF, Aquaron R, Pouget J (1986) Le déficit en AMP désaminase maladie ou syndrome. Rev Rhum 53:673-676
39. Kelemen J, Rice DR, Bradley WG, Munsat TL, DiMauro S, Hogan EL (1982) Familial myoadenylate deaminase deficiency and exertional myalgia. Neurology 32:857-863
40. Rubio JC, Martin MA, Bautista J, Campos Y, Segura D, Arenas J (1997) Association of genetically proven deficiencies of myophosphorylase and AMP deaminase: a second case of "double trouble". Neuromusc Disord 7:387-389
41. Figarella-Branger D, Pellissier JF, Perez-Castillo AM, Desnuelle C, Pouget J, Serratrice G (1991) Myopathie lentement progressive avec accumulation d'agrégats tubulaires. Rev Neurol (Paris) 147:586-594
42. Serratrice G, Pellissier JF (1997) Tubular aggregates myopathy in Congenital Muscular Dystrophy. In: Fukuyama Y, Osawa M, Saito K (eds) Elsevier, vol 1, pp 321-328
43. Lazaro J, Fenichel GM, Kilroy AW, Saito A, Fleischer S (1980) Cramps, muscle pain and tubular aggregate. Arch Neurol 37:715-717
44. Rosenberg NL, Neville HE, Ringel MD (1985) Tubular aggregates. Their association with neuromuscular diseases including the syndrome of myalgias/cramps. Arch Neurol 42:973-976
45. Niakan E, Harati Y, Danon MJ (1985) Tubular aggregates: their association with myalgia. J Neurol Neurosurg Psych 48:882-886
46. Brumback RA, Staton RD, Susag M (1981) Exercise-induced pain, stiffness, and tubular aggregation in skeletal muscle. J Neurol Neurosurg Psych 44:250-254
47. Salviati G, Pierobon-Bormioli S, Betto R (1985) Tubular aggregates; sarcoplasmic reticulum origin, calcium storage ability, and functional implications. Muscle Nerve 8:299-306

P-31 Magnetic Resonance Spectroscopy of Metabolic Abnormalities in Pathological Fatigue

D. Bendahan, P.J. Cozzone

Fatigue and/or muscle weakness is commonly reported by patients suffering from metabolic trouble, often of unknown etiology, affecting skeletal muscle.

The techniques used to measure dynamic metabolic events involved in muscular contraction are generally indirect in nature. Gas exchange and serum metabolites such as pyruvate, lactate and ammonia represent only averaged metabolic changes at the whole body level. Metabolism can also be directly investigated using tissue biopsies, a technique with significant limitations too. P-31 Magnetic Resonance Spectrometry (MRS) provides an opportunity to noninvasively and repeatably follow muscle metabolic changes *in vivo* in humans (in volumes of tissue large enough to be representative of the muscle as a whole) and has become over the years a tool of choice to investigate noninvasively muscular energetics. Since the first applications to animal models (isolated/perfused muscles) in 1974 [1] MRS technology has rapidly evolved with the development of RF-surface coils in 1980 [2] and the availability of high field wide-bore superconducting magnets. Metabolic explorations have been then quickly extended to *in situ* situations in animals and humans.

Over the past 15 years, a few research facilities have developed clinical programs with goal to document by P-31 MRS, the metabolic events taking place during muscular exercise in humans. Investigations cover both normal muscles subjected to exercise and diseased muscles. The CRMBM (Centre de Résonance Magnétique Biologique et Médicale) has developed in Marseille since 1987 a program to study muscular performance in healthy volunteers, athletes under training, and patients with various muscular pathologies.

In this paper, we survey our 10-year experience using P-31 MRS in human muscle in order to document metabolic abnormalities accounting for abnormal fatigue in patients suffering from various muscular disorders. A rest-exercise-recovery protocol was designed to be simple, short, informative and well-accepted by patients. From healthy volunteers to patients with abnormal muscular fatigue, the specific information brought by P-31 MRS to the understanding of both muscle physiology and physiopathology is described. The current results obtained with the standardized CRMBM protocol are reported. Specific applications to primary and secondary muscular disorders are presented with additional results on the objectivation ot the metabolic effects of palliative therapies.

Exploration of skeletal muscle by P-31 MRS: the method

P-31 MRS allows to observe directly and noninvasively the mobile phosphorylated metabolites present in muscular cells. This observation is both qualitative (identification of metabolites) and quantitative (determination of concentrations). The multiple advantages offered by P-31 MRS have already been detailed in previous reviews [3-5]. We will then limit the discussion to the main issues.

The phosphorus-31 nucleus possesses a spin 1/2 and its natural abundance is 100%. Its relatively high sensitivity is compatible with the recording of an adequate spectrum in less than 1 minute from a 1 mM solution of a phosphorylated compound. Phosphorylated molecules of biological interest resonate over a wide range of chemical shifts (ca. 40 ppm). This dispersion allows the discrimination of signals arising from compounds of similar chemical structures. Phosphorylated biomolecules bear usually a limited number of phosphorus atoms, hence the corresponding P-31 MRS spectra are often very simple and can be easily interpreted.

In Figure 1, we present the P-31 MRS spectrum recorded on the finger flexor muscles. Six resonances of variable intensities are easily identified. They correspond from left (downfield) to right (upfield) to the phosphorus atoms of phosphomonesters (PME), inorganic phosphate (Pi), phosphocreatine (PCr) and the triphosphate chain (Pα, Pβ and Pγ) of ATP. The PME signal is mostly ascribed to glucose 6-phosphate and, in some cases, AMP and IMP. The Pα signal of ATP displays occasionally an upfield shoulder corresponding to NAD^+/NADH. In some cases, signal corresponding to phosphodiesters (PDE) are observed between the Pi and PCr resonances. The inorganic phosphate, with a pKa of 6.7, exists at physiological pH under several ionized forms in rapid equilibrium, the relative concentrations of which vary as a function of pH and determine the position (chemical shift) of the Pi signal on the spectrum. With adequate reference curves (Fig. 1, insert), it is then possible to measure noninvasively the value of intracellular pH based on the Pi signal chemical shift. The measure of the intensity of each signal on the spectrum gives a quantitation of each phosphorylated compound.

One of the main contributions of P-31 MRS to muscle physiology is its ability to measure sequentially the variations of concentrations in PME, Pi, PCr, ATP as well as intracellular pH during muscle contrac-

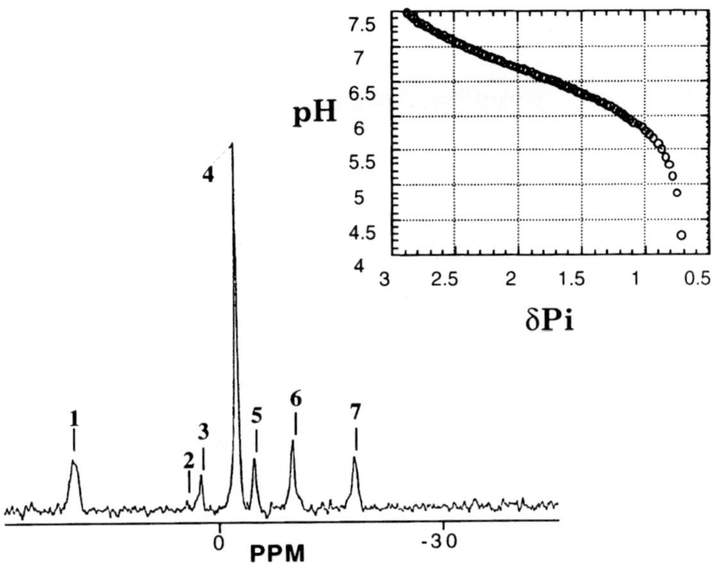

Fig. 1. P-31 MRS spectrum recorded on the finger flexor muscles of the forearm. All signals are assigned. The insert shows the reference curve used to determine the value of intracellular pH from the measurement of the chemical shift of the Pi signal. Assignments of resonances: (1) reference compound (phenylphosphonic acid); (2) phosphomonoesters; (3) Pi; (4) PCr; (5) Pγ of ATP; (6) Pα of ATP, and (7) Pβ of ATP

tion and recovery. Biochemical reactions taking place in the myocyte are then directly reflected by specific perturbations on the spectrum. These reactions are summarized in Fig. 2. The experimental set-up which has been developed is shown in Fig. 3. It allows to study the metabolic events occurring in the finger flexor muscles of the forearm when the fingers are subjected to a rest-exercise-recovery protocol. This ergometric apparatus has been built to fit in the horizontal bore (diameter: 30 cm) of the 4.7 T horizontal magnet of a Bruker Biospec spectrometer. Spatial selection is achieved with a surface coil which acts as a "metabolic stethoscope" for the muscles of interest. A typical series of spectra, recorded at 15 sec intervals throughout the protocol is displayed in Fig. 4. During contraction, the free energy stored in ATP (high phosphate-group transfer potential) is converted into mechanical energy via the interaction of two muscle proteins, actin and myosin. Muscle work is linked to hydrolysis of ATP. However, in normal subjects the intracellular concentration of ATP remains unchanged during exercise (Fig. 5) due to an adequate balance between energy demand and supply. Net intracellular acidosis due to lactic acid production from glycogenolysis is reflected on the spectrum by an upfield shift of the Pi signal.

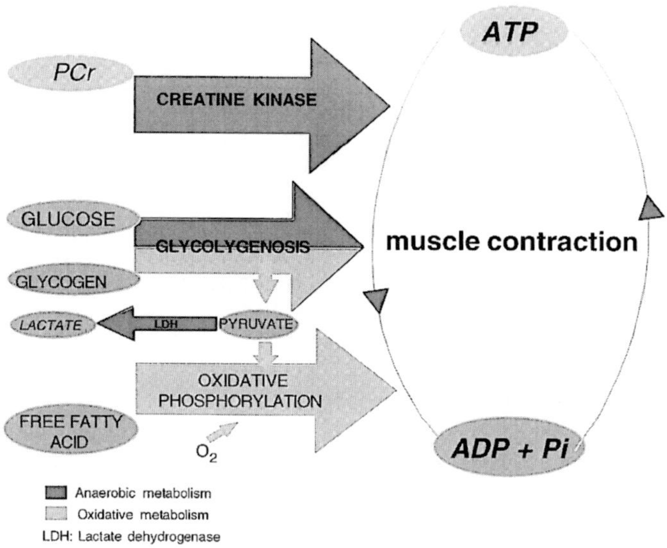

Fig. 2. Metabolic pathways of muscle exercise. During the first seconds of exercise, the breakdown of PCr maintains ATP homeostasis. When exercise is prolonged, other reactions meet the energy demand. Metabolites directly observed and assayed by P-31 MRS are shown in italics

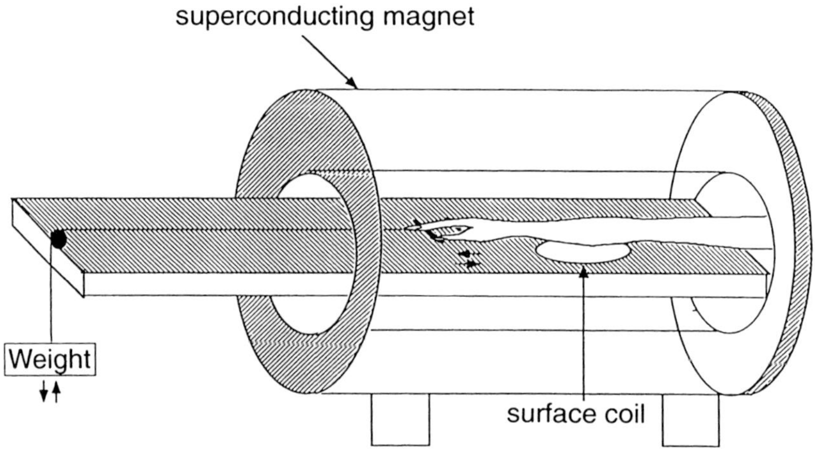

Fig. 3. Experimental set-up developed to record P-31 MRS spectra on the finger flexor muscles at rest, exercise and during recovery. The apparatus, designed and built at CRMBM (Marseille), is interfaced with the 4.7 T supraconducting magnet of a Bruker Biospec 47/30 system. Signals are detected by a surface coil placed on the forearm. The intensity of the finger flexions is calibrated with a pulley-weight device

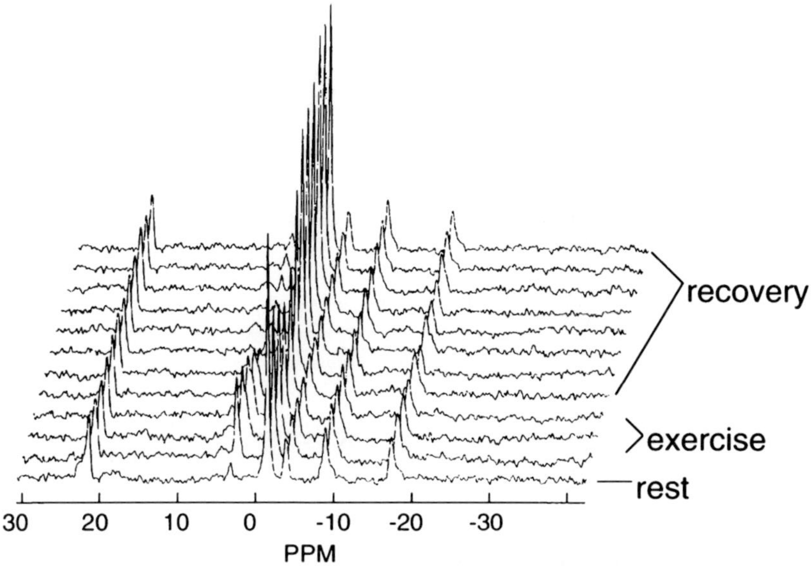

Fig. 4. Series of P-31 MRS spectra recorded on the finger flexor muscle of a healthy volunteer during the standardized protocol. Assignments are as indicated on Fig. 1

Exploration of abnormal muscle fatigue by P-31 MRS

Fatigue is classically defined as the inability to sustain a required mechanical force. The general phenomenon of fatigue characterized as a transient and reversible process which regulates and protects the physiology of an exercising muscle has been extensively studied with 31-P MRS and is not the subject of this paper. We will focus instead on the pathological fatigue with asthenia, chronic fatigue syndrome or a fatigue associated to a metabolic disorder, expressing a durable pathological situation. Before characterizing any abnormal metabolic status related to fatigue one should establish what can be considered as "normality".

Definition by P-31 MRS of a normal metabolic response to a standardized rest-exercise-recovery protocol

Numerous studies have tried to define the normal metabolic response to exercise of a healthy volunteer using several experimental protocols having different duration and intensity. We have developed at the CRMBM in Marseille, a protocol to study the metabolism of finger flexors muscles with a Bruker Biospec spectrometer, operating at 4.7 T and equipped with a horizontal superconducting magnet (bore diameter: 30 cm). Spatial localization and signal recording are achieved with a double tuned (phosphorus, proton) surface coil. Field homogeneity is adjusted during rest on

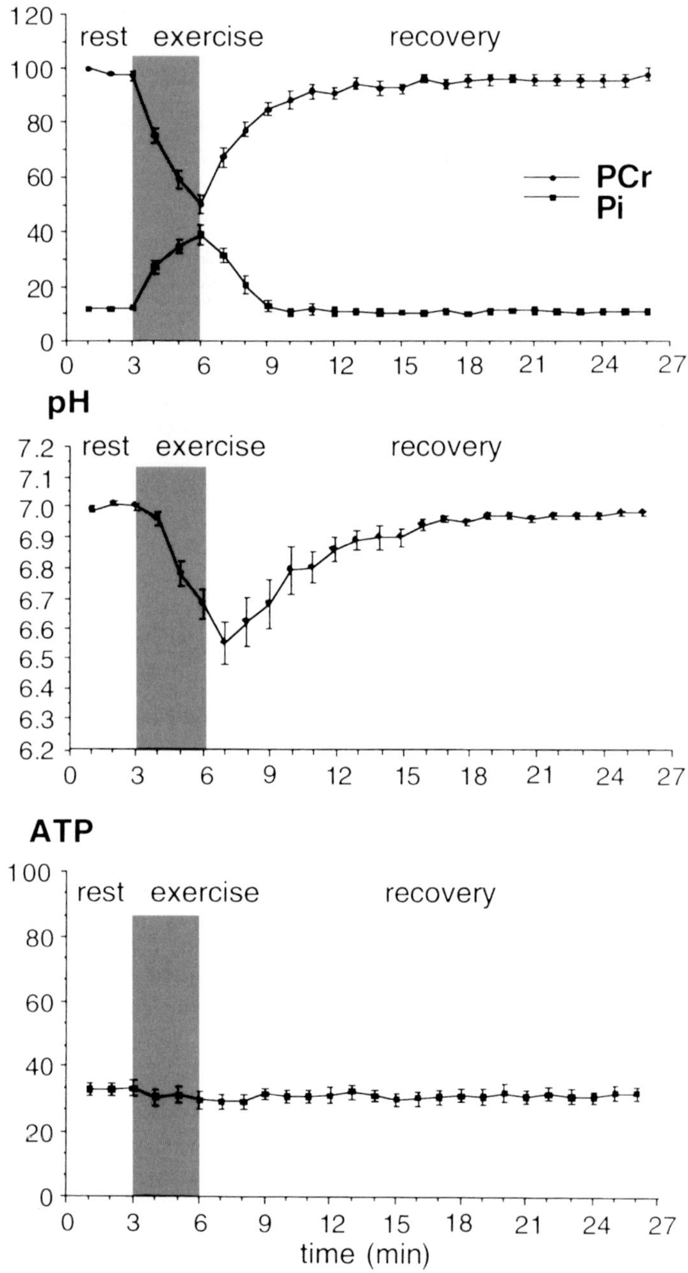

Fig. 5. Mean metabolic profiles recorded by P-31 MRS on 25 healthy volunteers subjected to the rest-exercise-recovery protocol. The shaded areas correspond to the exercise period. Homeostasis of ATP is maintained. The slight acidosis at the onset of recovery is due to PCr resynthesis

the intracellular water signal at 200.14 MHz. Fully relaxed P-31 NMR spectra are recorded at rest in ca. 3 min (8 scans) following 55 ms radiofrequency pulses applied every 20s. Then, sequential P-31 NMR spectra are recorded every 15 seconds (8 scans) following 120 ms radiofrequency pulses applied every 1.9s. Exercise consists of finger flexions lifting a 6 Kg weight every 1.5s for 3 min. The high magnetic field (4.7 T) allows to obtain in several seconds a spectrum of adequate signal-to-noise ratio.

The analysis of spectra recorded on a cohort of about 100 healthy volunteers shows the existence of large interindividual heterogeneity. However, we have established criteria of metabolic normality encompassing this natural diversity. These invariant parameters, which are a prerequisite to any investigation of pathological situations, are expressed as linear relationships between energy consumption and acidosis during exercice; they quantitate the dependence of recovery kinetics upon end-of-exercise pH [6, 7]. The use of these correlations offers the distinct advantage to capture the information contained in interindividual heterogeneity.

We have originally described the evolution of sugar monophosphates throughout the rest-exercise-recovery protocol. It brings new information on the activation mechanisms of glycolysis during exercise [8]. We have shown that the transient disappearance of Pi during the early phase of recovery can be considered as an index of glycogen phosphorylase activity. To summarize, we have defined on the basis of P-31 MRS, criteria of metabolic normality at rest, during muscle exercise and subsequent recovery. Those encompass the interindividual variability. The total duration of the protocol is ca. 30 min which makes it easily accepted by volunteers and more so by patients with limited tolerance to exercise.

Pathological states of fatigue

Metabolic myopathies are often accompanied by an intolerance to muscle effort and an extreme fatigue. P-31 MRS is used to study muscle energetics in a number of disorders affecting skeletal muscle directly or indirectly.

Mitochondrial diseases

The metabolic profile recorded on a 19 year-old patient with Kearns-Sayre's disease is presented in Fig. 6. The series of spectra collected during the rest-exercise-recovery protocol indicate abnormalities characterizing a mitochondrial disorder. The PCr/Pi ratio is low indicating an increased level of free ADP at rest, and the half-time of PCr recovery is increased. This slower rate of PCr resynthesis reflects an impaired mitochondrial function. During exercise, acidosis is limited, i.e. lower than expected according to the linear relationship between pH and PCr/Pi both measured at end of exercise. In this type of pathology, lactic acidosis (reflecting en-

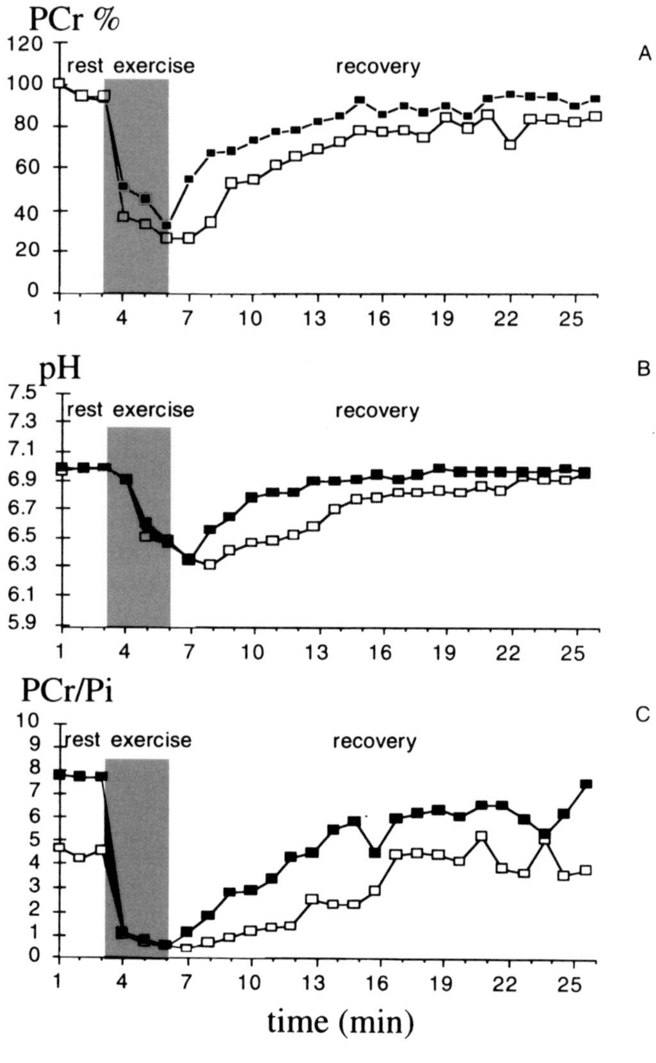

Fig. 6. Therapeutic follow-up by P-31 MRS of a patient with Kearns-Sayre's disease, treated by coenzyme Q10. The profiles of PCr (A), pH (B) and PCr/Pi (C) evolve favorably after 10 months of treatment. Note the acceleration in the initial rates of recovery for the 3 parameters.
□ before treatment, ■ after treatment.
This patient is monitored in collaboration with the Clinic of Neuromuscular Diseases at the Timone Hospital in Marseille (Prof. G. Serratrice, J. Pouget and C. Desnuelle); the Department of Neuropathology (Prof. J.F. Pellissier and Dr. D. Figarella-Branger), and the Department of Metabolic Explorations (Prof. Vanuxem, Sainte Marguerite Hospital). A group of 15 patients is currently followed longitudinally according to the same protocol

hanced glycolytic activity) is often observed. In the present case, the limited acidosis indicates an adaptative mechanism which counteracts the expected additional acidosis. This adaptative mechanism has been further confirmed by measurements of proton efflux which is, most of the times, abnormally fast for patients with mitochondrial myopathies [9, 10]. Also, maximal aerobic capacity, calculated from ADP concentration and kinetics of PCr recovery, is sometimes reduced for patients with oxidative disorders [10]. For this kind of pathologies, it should be kept in mind that genetic and phenotypic expressions are largely heterogeneous. Metabolic abnormalities recorded for these patients clearly reflect this variability. At the present time, relationships between clinical expression and metabolic abnormalities remain obscure. It is noteworthy that limited acidosis and slower kinetics of PCr and/or ADP recovery are the most commonly observed abnormalities in oxidative disorders [11-13]. Large breakdown of PCr during exercise and a slowdown of post exercise recovery of Pi and PCr levels [14] have been reported for patients with malate-aspartate shuttle deficiency. In this pathology, transfer of reducing equivalents from cytosol to mitochondria is impaired.

In some other cases, for which skeletal muscle is secondarily affected, metabolic abnormalities suggesting an oxidative disorder have also been described. Patients with hypothyroidism display slow kinetics of PCr recovery associated with reduced ratio of PCr/Pi at rest [15]. Limited acidosis and faster proton efflux have been reported for patients with primary hypertension [16, 17]. These abnormalities have been related to a hyperactivity of the sodium-proton antiport. Interestingly, for these patients, ADP recovery kinetics is normal [16].

Glycogenoses

Similarly, muscle fatigue and decreased performance in patients with glycolytic disorders are accompanied by alterations of spectra. The metabolic profiles of patients with a deficiency in glycogen phosphorylase (Mc Ardle's disease) reveals clear anomalies such as the lack of acidosis during exercise due to the inhibited glycogenolysis (Fig. 7). In fact, a weak alkalosis is often observed due to the consumption of H^+ in relation to excessive breakdown of PCr. An overproduction of Pi during exercise is also apparent. The Pi signal remains visible throughout recovery, bringing additional information to the initial observations made by Ross et al. [18] and Duboc et al. [19]. Moreover, comparison with the criteria of normality shows a slower recovery of the PCr/Pi ratio, illustrating an associated oxidative disorder because of a shortage in substrates, in perfect agreement with other metabolic measurements such as maximum oxygen consumption [20]. In other glycolytic pathologies, the overproduction of PME and their remanence during recovery suggest more distal disturbances of glycogenolysis, such as phosphofructokinase deficiency [19, 21].

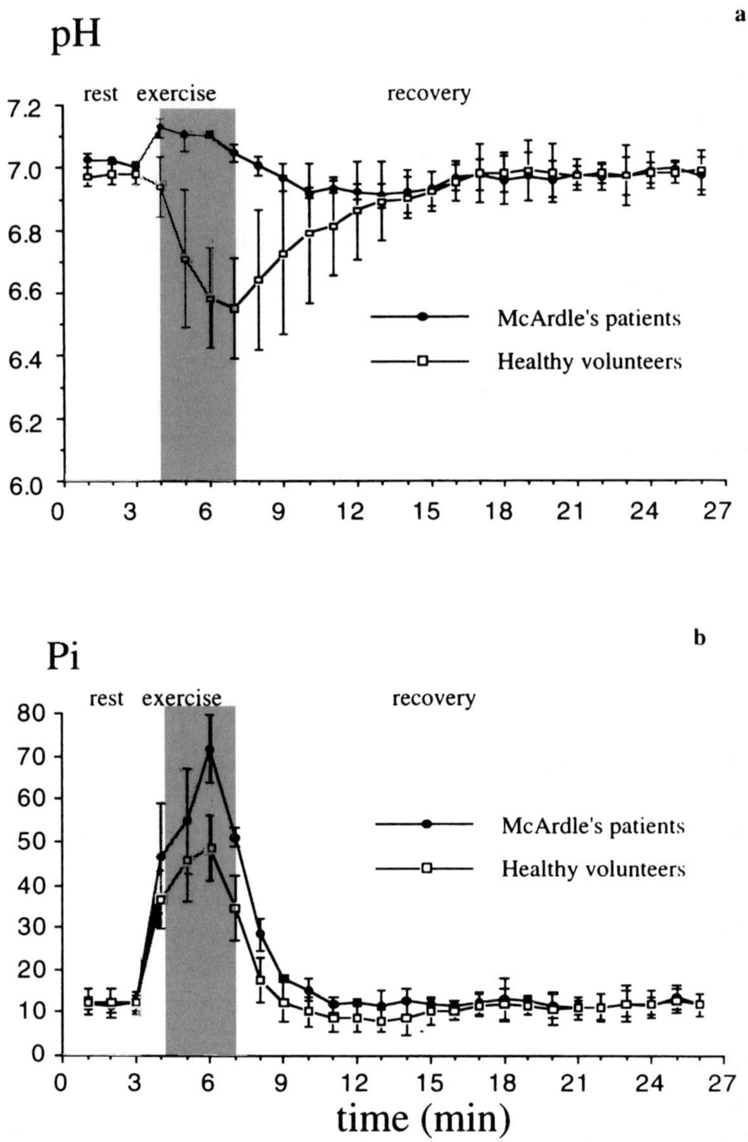

Fig. 7. Mean metabolic profiles recorded by P-31 MRS on 3 patients with Mc Ardle's disease during a rest-exercice-recovery protocol. The shaded areas correspond to the exercice period. • Patients with Mc Ardle's disease; □ healthy volunteers (n = 14). Note the lack of acidosis during exercice and the overproduction of Pi for the 3 patients

Glycogenolytic hyperactivity

P-31 MRS has also been used to document abnormal fatigue associated with other disorders affecting skeletal muscle directly or indirectly. It is noteworthy that hyperacidosis coupled to excessive breakdown of PCr has been reported for various pathological states including calcium ATP-ase deficiency [22], post-viral syndrome [23] and hyperthyroidism [24, 25]. Similar abnormalities have been reported for a patient with tubular aggregates, complaining from severe intolerance to effort [26]. It has to be underlined that the similarity of metabolic abnormalities reported for different clinical situations considerably reduced the specificity of P-31 MRS as a diagnostic tool. However, the understanding of metabolic abnormalities in terms of altered muscle energetics can be facilitated through a quantitative analysis.

Diagnosis of malignant hyperthermia

Diagnosis of Malignant Hyperthermia Susceptibility (MHS), a pharmacogenetic disorder, is usually performed on muscle biopsy specimen submitted to *in vitro* contracture tests. As an alternative noninvasive approach, several research groups have screened MHS subjects with P-31 MRS. The first purpose of these studies has been to detect metabolic abnormalities that could account for the abnormal muscular reaction during anesthesia with fluorinated compounds [27-31]. Metabolic abnormalities have been detected at rest [29, 30], during exercise [27, 28, 31, 32] or during the post-exercise recovery phase [28, 29]. On the other hand, diagnostic strategies, based on the selection of parameters recorded at rest [33, 34] or during a standardized rest-exercise-recovery protocol [32], have been reported. These strategies provided sensitivity and specificity ranging from 95 [33, 34] to 100% [32], thereby demonstrating the diagnostic capacity of MRS for malignant hyperthermia susceptibility. These results should be confirmed through prospective studies of larger group of patients.

Quantitative approach of pathological fatigue states

More recently, P-31 MRS has been used to quantitatively document abnormal muscle fatigue in various pathological conditions. Briefly, the combined analysis of metabolic changes recorded during and after exercise has been used to determine contributions of glycolytic and oxidative processes to energy production, proton efflux and maximum aerobic capacity according to the original work of the Oxford's group [35]. For mitochondrial myopathies, this quantitative approach has further confirmed the upregulation of proton efflux leading to limited pH changes and the heterogeneity of this multisystemic family of disorders [35]. Reduction of maximum aerobic capacity has been reported with no clear correlation with clinical findings

[10]. Similar anomalies have been reported for patients with maternally inherited diabetes and deafness secondary to point mutation of mtDNA [36, 37]. Besides the upregulation of proton efflux, we have reported increased energy cost of contraction which could be related either to the presence of ragged-red fibers (often described in oxidative disorders) or to hypermetabolism described as early as 1962 for a patient with Luft's syndrome [38]. Weakness and fatigue associated with chronic renal failure has been largely documented using the same quantitative approach [39, 40]. In dialysed patients, muscle atrophy and/or alteration of contractile efficiency and/or defective oxidative function can account for substantial abnormalities of muscle energetics. Oxidative defect could be a result of anemia [40]. In patients with claudication due to peripheral vascular disease, reduced blood flow alters proton efflux and oxidative ATP generation [41]. However, increased energy cost of contraction (twice the normal value) is probably accounted for by reduced metabolic efficiency in addition to reduced muscle blood flow. Finally, in patients with congestive heart failure, decreased oxidative capacity and metabolic efficiency both contribute to increased phosphocreatine depletion and intracellular acidosis during exercise [42].

Palliative therapies

A unique feature of P-31 MRS is its ability to monitor noninvasively the possible effects of a palliative therapy. A number of results have been reported on the improvement of muscle weakness following various therapies. Most of them pertain to patients with defective oxidative function. We reported, in agreement with some other groups, that long term treatment with coenzyme Q_{10} improves the post-exercise recovery process while being accompanied by an increase in overall muscle performance and tolerance to exercise [12, 43, 44]. One study has identified responders and non responders to coenzyme Q_{10} therapy on the basis of P-31 MRS measurements while a positive effect could not have been predicted on a simple basis [44]. On the other hand, short-term treatment with Coenzyme Q_{10} and multiple vitamins has proved ineffective in treatment of mitochondrial disease [45]. Short-term treatment with dichloroacetate, a drug able to reduce lactic acid accumulation by stimulation of its oxidation to acetyl-CoA, has been shown to improve muscle performance, when combined with aerobic training, in one subject with cytochrome oxidase deficiency [46]. The results of some other P-31 MRS trials have been reported for patients with intermittent claudication and acute asthenia [47, 48]. Patients with intermittent claudication displayed a reduction of acidosis during exercise and increased oxidative ATP generation during recovery after a short-term propionyl-L-carnitine treatment. This positive metabolic effect is only significant for patients with increased maximum walking distance after therapy. Finally, citrulline malate induces a reduction of acidosis and improves the kinetics of PCr post-exercise recovery for patients with acute asthenia [47]. These changes are coupled to trends indicating a reduction of anaerobic and increased aerobic ATP production [47].

Conclusion

The noninvasiveness of P-31 MRS is a major competitive advantage of this technique in the study of neuromuscular diseases. Conventional techniques (biopsy, EMG, etc.) are more invasive and generally less accepted by patients. The ergometric bicycle test is often too strenuous to be feasible in the presence of incapacitating conditions. Comparatively, P-31 MRS appears as a tool of choice to investigate muscle energetics. However, although MRS sensitivity is high, as shown above, in the detection of metabolic deviations, its specificity remains limited. Obviously, specific glycolytic disorders and mitochondriopathies cannot be mistaken and the diagnostic value of P-31 MRS is significant in a number of muscle pathologies. However, in many cases, the metabolic heterogeneity of muscle disorders makes it difficult to pose an unquestionable diagnosis. In these cases, the results of P-31 MRS are more difficult to interpret in specific terms and they have to be combined on a pluridisciplinary basis with other investigative methods. In all cases, when the diagnosis is reached, P-31 MRS emerges as the method of election to ensure a noninvasive follow-up of disease evolution, to establish the effectiveness of therapy and to screen families of patients with hereditary forms of muscle diseases.

Acknowledgements

Investigation of muscle energetics at CRMBM is performed in collaboration with J.P. Mattéi, M. Roussel, S. Confort-Gouny, Y. Lefur and G. Kozak-Ribbens. Physiopathological applications are conducted in collaboration with Professors J. Pouget (Service des Maladies du Système Nerveux et de l'Appareil Locomoteur, CHU de la Timone, Marseille), C. Desnuelle (Service de Neurologie, Hôpital Pasteur, Nice), J.F. Pellissier and D. Figarella-Branger (Service d'Anatomie Pathologique et de Neuropathologie, CHU de la Timone, Marseille), J.L. Gastaut (Service de Neurologie, Hôpital de Sainte Marguerite, Marseille), H. Roux (Service de Rhumatologie, Hôpital de la Conception, Marseille), M. Aubert (Service d'Anesthésie Réanimation, HIA Laveran, Marseille), and D. Vanuxem (Laboratoire d'Explorations Fonctionnelles et Respiratoires, Hôpital de Sainte Marguerite, Marseille). This research was supported by grants from the Centre National de la Recherche Scientifique (UMR n° 6612), the Association Française contre les Myopathies (AFM), the Assistance Publique – Hôpitaux de Marseille AP-HM) and the Programme Hospitalier de Recherche Clinique (PHRC).

References

1. Hoult DI, Busby JW, Gadian DG, Radda GK, Richards RE, Seeley PJ (1974) Observation of tissue metabolites using 31-P nuclear magnetic resonance. Nature 252:285-287

2. Ackerman JH, Groove TH, Wong G, Gadian DG, Radda GK (1980) Mapping of metabolites in whole animals by 31-P NMR using surface coil. Natur 283:167-170
3. Bernard M, Canioni P, Cozzone PJ (1983) Étude du métabolisme cellulaire in vivo par résonance magnétique nucléaire du phosphore-31. 1. Biochimie 65:449-470
4. Meyer RA, Kushmerick Mj, Brown TR (1982) Application of 31-P NMR Spectroscopy to the study of striated muscle metabolism. Am J Physiol 242:C1-C11
5. Sapega AA, Sokolow DP, Graham TJ, Chance B (1987) Phosphorus nuclear magnetic resonance: a non invasive technique for the study of muscle bioenergetics during exercise. Med Sci Sports Exerc 19:410-420
6. Bendahan D, Confort-Gouny S, Kozak-Reiss G, Cozzone PJ (1990) Heterogeneity of metabolic response to exercise in humans: new criteria of invariance defined by in vivo phosphorus NMR spectroscopy. FEBS Lett 272:155-158
7. Bendahan D, Kozak-Reiss G, Confort-Gouny S, Cozzone PJ (1991) *In vivo* 31-P NMR determination of metabolic parameters characterizing the bioenergetic response to muscular exercise in humans. Proceedings of the Tenth Annual Scientific Meeting. Soc Magn Reson Med 2:545
8. Bendahan D, Confort-Gouny S, Kozak-Reiss G, Cozzone PJ (1990) Pi trapping in glycogenolytic pathway can explain transient Pi disappearance during recovery from muscular exercise : a P-31 NMR study in the human. FEBS Letters 269:402-405
9. Kemp GJ, Thompson CH, Taylor DJ, Hands LJ, Rajagopalan B, Radda GK (1993) Quantitative analysis of defects of mitochondrial oxidation during recovery from exercise. Biochem Soc Trans 21:176S
10. Taylor DJ, Kemp GJ, Radda GK (1994) Bioenergetics of skeletal muscle in mitochondrial myopathy. J Neurol Sci 127:198-206
11. Arnold DL, Taylor Dj, Radda GK (1985) Investigation of human mitochondrial myopathies by phosphorus magnetic resonance spectroscopy. Ann Neurol 18:189-196
12. Bendahan D, Desnuelle C, Vanuxem D, Confort-Gouny S, Figarella-Branger D, Pellissier JF, Kozak-Ribbens G, Pouget J, Serratrice G, Cozzone PJ (1992) 31-P NMR spectroscopy and ergometer exercice test as evidence for muscle oxidative performance improvement with coenzyme Q in mitochondrial myopathies. Neurology 42:1203-1208
13. Matthews P, Allaire C, Shoubridge E, Karpati G, Carpenter S, Arnold D (1991) *In vivo* Magnetic resonance spectroscopy in the clinical investigation of mitochondrial disease. Neurology 41:114-120
14. Hayes Dj, Taylor DJ, Bore PJ, Hilton-Jones D, Arnold DL, Squier MV, Gent A, Radda GK (1987) An unusual metabolic myopathy: a malate-aspartate shuttle defect. J Neurol Sc 82:27-39
15. Kaminsky P, Robin-Lherbier B, Brunotte F, Escanye JM, Walker P, Klein M, Robert J, Duc M (1992) Energetic metabolism in hypothyroid skeletal muscle as studied by phosphorus magnetic resonance spectroscopy. J Clin Endocrinol Metab 74:124-129
16. Dudley CR, Taylor DJ, NG LL, Kemp GJ, Ratcliffe Pj, Radda GK, Ledingham JG (1990) Evidence for abnormal Na^+/H^+ antiport activity detected by phosphorus nuclear magnetic resonance spectroscopy in exercising skeletal muscle of patients with essential hypertension. Clin Sci 79:491-497
17. Ronquist G, Soussi B, Frithz G, Schersten T, Waldenstrom A (1995) Disturbed energy balance in skeletal muscle of patients with untreated primary hypertension. J Intern Med 238:167-74
18. Ross BD, Radda GK, Gadian DG, Rocker G, Esiri M, Falconer-Smith J (1981) Examination of a case of suspected McArdle's syndrome by 31-P nuclear magnetic resonance. N Engl J Med 28:1338-1342
19. Duboc D, Jehenson P, Tran-Dinh S, Marsac C, Syrota A, Fardeau M (1987) Phosphorus NMR spectroscopy study of muscular enzyme deficiencies involving glycogenolysis and glycolyses. Neurology 37:663-671
20. Bendahan D, Confort-Gouny S, Kozak-Ribbens G, Cozzone PJ (1992) P-31 NMR characterization of the metabolic anomalies associated with the lack of glycogen

phosphorylase activity in human forearm muscle. Biochem Biophys Res Commun 185:1621

21. Argov Z, Bank Wj, Maris J, Leigh JS, Chance B (1987) Muscle energy metabolism in human phosphofructokinase deficiency as recorded by 31-P nuclear magnetic resonance spectroscopy. Ann Neurol 22:46-51
22. Taylor DJ, Brosnan MJ, Arnold DL, Bore PJ, Styles P, Walton J, Radda GK (1988) Ca^{2+} ATPase deficiency in a patient with an exertional muscle pain syndrome. J Neurol Neuros Psychiatry 51:1425-1433
23. Arnold DL, Bore Pj, Styles Pj, Radda GK, Taylor Dj (1984) Excessive intracellular acidosis of skeletal muscle on exercise in a patient with a post-viral exhaustion/fatigue syndrome: A P-31 Nuclear Magnetic Resonance Study. Lancet 8391:1367-1369
24. Argov Z, Renshaw PF, Boden B, Winokur A, Bank Wj (1988) Effects of thyroid hormones on skeletal muscle bioenergetics: in vivo phosphorus-31 magnetic resonance spectroscopy study of human and rats. J Clin Invest 81:1695-1701
25. Kaminsky P, Robin Lherbier B, Walker P, Brunotte F, Escanye JM, Klein M, Forrett MC, Robert J, Duc M (1991) Muscle bioenergetic impairment in hyperthyroid man: a study by 31-P NMR spectroscopy. Acta Endocrinol Copenh 124:271-277
26. Bendahan D, Pouget J, Pellissier JF, Figarella-Branger D, Cozzone PJ (1996) Magnetic resonance spectroscopy and histological study of tubular aggregates in a familial myopathy. Journal of the Neurological Sciences 139:149-155
27. Kozak-Reiss G, Gascard JP, Redouane-Bénichou K (1986) Malignant hyperthermia: diagnostic means and the possibilities opened by NMR spectroscopy. Ann Fr Anesth Réanim 5:584-589
28. Monsieurs K, Heytens L, Kloeck C, Martin Jj, Wuyts F, Bossaert L (1997) Slower recovery of muscle phosphocreatine in malignant hyperthermia-susceptible individuals assessed by P-31-MR spectroscopy. J Neurol 244:651-656
29. Olgin J, Argov Z, Rosenberg H, Tuchler M, Chance B (1988) Non-invasive evaluation of malignant hyperthermie susceptibility with phosphorus nuclear magnetic resonance spectroscopy. Anesthesiology 68:507-513
30. Payen JF, Bourdon L, Mezin P, Jacquot C, Le Bas JF, Stieglitz P, Benabid AL (1991) Susceptibility to malignant hyperthermia detected non invasively. Lancet 337:1550-1551
31. Webster DW, Thompson RT, Gravelle DR, Laschuk MJ, Driedger AA (1990) Metabolic response to exercise in Malignant Hyperthermia-sensitive patients measured by 31-P magnetic resonance spectroscopy. Magn Res Med 15:81-89
32. Bendahan D, Kozak-Ribbens G, Rodet L, Confort-Gouny S, Cozzone PJ (1998) (31)phosphorus magnetic resonance spectroscopy characterization of muscular metabolic anomalies in patients with malignant hyperthermia- application to diagnosis. Anesthesiology 88:96-107
33. Olgin J, Rosenberg H, Allen G, Seestedt R, Chance B (1991) A blinded comparison of non invasive, in vivo phosphorus nuclear magnetic resonance spectroscopy and the in vitro halothane/caffeine contracture test in the evaluation of malignant hyperthermia suceptibility. Anesth Analg 72:36-47
34. Payen JF, Bosson JL, Stieglitz P (1996) Prospective evaluation of a nuclear magnetic resonance score in the screening of malignant hyperthermia susceptibility. [Initial results] Ann Fr Anesth Reanim 15:47-48
35. Kemp GJ, Taylor DJ, Thompson CH, Hands LJ, Rajagopalan B, Styles P, Radda GK (1993) Quantitative analysis by 31-P magnetic resonance spectroscopy of abnormal mitochondrial oxidation in skeletal muscle during recovery from exercise. NMR Biomed 6:302-10
36. Vialettes B, Paquis-Fluckinger V, Pelissier JF, Bendahan D, Narbonne H, Silvestre-Aillaud P, Montfort MF, Righini-Chossegros M, Pouget J, Cozzone P, Desnuelle C (1997) Phenotypic expression of diabetes secondary to a T 14709 C mutation of mitochondrial DNA. Comparison with the MIDD syndrome (A 3243 G mutation). Diabetes Care 20:1731-1737

37. Vialettes B, Paquis-Fluckinger V, Silvestre-Aillaud P, Bendahan D, Pelissier JF, Etchary-Bouyx F, Raccah D, Gin H, Guillausseau PJ, Vanuxem D, Orsini F, Pouget J, Cozzone P, Desnuelle C (1995) Extra-pancreatic manifestations in diabetes secondary to mitochondrial DNA point mutation within the tRNA(Leu(UUR)) gene. Diabetes Care 18:1023-1028
38. Palmieri G, Ernster L, Afzelius B (1962) A case of severe hypermetabolism of nonthyrois origin with a defect in the maintenance of mitochondrial respiratory control : a correlated clinical, biochemical and morphological study. J Clin Invest 41:1776-1804
39. Kemp GJ, Thompson CH, Taylor DJ, Radda GK (1995) ATP production and mechanical work in exercising skeletal muscle: a theoretical analysis applied to 31-P magnetic resonance spectroscopy studies of dialyzed uremic patients. Magn Reson Med 33:601-609
40. Thompson RT, Muirhead N, Marsh GD, Gravelle D, Potwarka JJ, Driedger AA (1996) Effect of anaemia correction on skeletal muscle metabolism in patients with end- stage renal disease: 31-P magnetic resonance spectroscopy assessment. Nephron 73:436-441
41. Kemp GJ, Hands LJ, Ramaswami G, Taylor DJ, Nicolaides A, Amato A, Radda GK (1995) Calf muscle mitochondrial and glycogenolytic ATP synthesis in patients with claudication due to peripheral vascular disease analysed using 31-P magnetic resonance spectroscopy. Clin Sci Colch 89:581-90
42. Kemp GJ, Thompson CH, Stratton JR, Brunotte F, Conway M, Adamopoulos S, Arnolda L, Radda GK, Rajagopalan B (1996) Abnormalities in exercising skeletal muscle in congestive heart failure can be explained in terms of decreased mitochondrial ATP synthesis, reduced metabolic efficiency, and increased glycogenolysis. Heart 76:35-41
43. Barbiroli Bl, Frassineti C, Martinelli P, Lotti S, Lodi R, Cortelli P, Montagna P (1997) Coenzyme Q_{10} improves mitochondrial respiration in patients with mitochondrial cytopathies – an *in vivo* study on brain and skeletal muscle by phosphorus magnetic resonance spectroscopy. Cel Mol Biol 43:741-749
44. Gold R, Seibel P, Reinelt G, Schindler R, Landwehr P, Beck A, Reichmann H (1990) Phosphorus magnetic resonance spectroscopy in the evaluation of mitochondrial myopathies: results of a 6-month therapy study with coenzyme Q. Eur Neurol 36:191-196
45. Matthews P, Ford B, Dandurand R, Eidelman D, O'Connor D, Sherwin A, Karpati G, Anderman F, Arnold D (1993) Coenzyme Q10 with multiple vitamins is generally ineffective in treatment of mitochondrial disease. Neurology 43:884-890
46. Taivassalo T, Matthews PM, De SN, Sripathi N, Genge A, Karpati G, Arnold DL (1996) Combined aerobic training and dichloroacetate improve exercise capacity and indices of aerobic metabolism in muscle cytochrome oxidase deficiency. Neurology 47:52934
47. Bendahan D, Mattei JP, Confort-Gouny S, Leguern ME, Musial C, Alquier C, Cozzone PJ (1997) 31-P MRS analysis of muscle metabolic effects of citrulline malate in humans. MAGMA 5:139
48. Taylor DJ, Amato A, Hands LJ, Kemp GJ, Ramaswami G, Nicolaides A, Radda GK (1996) Changes in energy metabolism of calf muscle in patients with intermittent claudication assessed by 31-P magnetic resonance spectroscopy: a phase II open study.Vasc Med 1:241-245

New Aspects of Metabolic Myopathies

S. DiMauro, C. Bruno, A.L. Andreu, G.M. Hadjigeorgiou, S. Shanske

There are two major clinical presentations of the metabolic myopathies: one is fixed, usually slowly progressive weakness; the other corresponds exactly to the main theme of this symposium, exercise intolerance, muscle contracture and pain. We will confine our review to recent developments in this second group of disorders. Although acute muscle breakdown (rhabdomyolysis) with myoglobinuria is not explicitly mentioned in the title of the book, in fact most, if not all, of the metabolic myopathies under consideration can cause myoglobinuria and are listed in Table 1. When one compares this table to other lists of the heritable myoglobinurias published as recently as 1992 [1], the striking difference is in the number of defects of the respiratory chain that have been added recently. This is largely due to the increased attention devoted to the respiratory chain by clinical scientists since 1988, when mutations in mitochondrial DNA (mtDNA) were first described [2, 3], and it is hardly surprising when one considers that aerobic metabolism is the major source of energy for muscle contraction [4].

Glycogen storage diseases

Even in this well trodden area of muscle metabolism, a new disease has been recently added to the list, *aldolase deficiency* [5]. A 4 1/2-year-old boy was admitted to the hospital because of exercise intolerance and weakness two days after a febrile illness. No pigmenturia was described, but myoglobin was detected in the urine and serum CK was 2,620 U/L (normal, less than 60). The child also had evidence of hemolytic anemia, and was slightly delayed in his motor development and language acquisition. On follow-up, there were several other episodes of exercise intolerance, weakness, and serum CK elevations associated with intercurrent febrile illnesses. Muscle biopsy was normal morphologically, but showed a profound defect of aldolase activity (about 10% of normal) wheras all other glycolytic enzymes had normal activities. Molecular genetic analysis revealed a Glu206Lys mutation in a highly conserved region of the aldolase molecule. This case is instructive because the patient had rather subtle symptoms triggered by fever rather than exercise, no obvious pigmenturia or massive serum CK elevation, and normal muscle biopsy. Aldolase determination should be added to the battery of biochemical tests to be performed in patients with intermittent exercise intolerance, weakness, and hyper-CKemia.

Table 1. Metabolic causes of exercise intolerance and myoglobinuria

Glycogen storage diseases
1. Phosphorylase b kinase
2. Myophosphorylase (McArdle disease)
3. Phosphofructokinase (PFK-M) (Tarui disease)
4. Aldolase
5. Phosphoglycerate kinase (PGK)
6. Phosphoglycerate mutase (PGAM-M)
7. Lactate dehydrogenase (LDH-A)
Disorders of lipid metabolism
1. Carnitine palmitoyltransferase II (CPT II)
2. Very long-chain acyl-CoA dehydrogenase (VLCAD)
3. Long-chain acyl-CoA dehydrogenase (LCAD)
4. Long-chain 3-hydroxyacyl-CoA dehydrogenase (LCHAD)
5. Short-chain 3-hydroxyacyl-CoA dehydrogenase (SCHAD)
6. Trifunctional protein (TP)
Defects of pyruvate utilization
Lipoamide dehydrogenase (LAD)
Defects of the pentose phosphate pathway
Glucose-6-phosphate dehydrogenase (G6PD)
Respiratory chain defects
1. Coenzyme Q10
2. Complex II and aconitase
2. Complex III (mutations in cytochrome b gene)
3. Complex IV (COX)

The discovery in 1993 [6] of three distinct mutations in the myophosphorylase gene in patients with *myophosphorylase deficiency (McArdle disease)* was an accurate presage of the remarkable genetic heterogeneity that prevails in this disease as well as in other glycogenoses. As of 1997, 11 mutations had been described in McArdle disease [7] in different ethnic groups, although one of them, Arg49Stop, was by far the most frequent, at least in predominantly Anglosaxon populations. Bartran et al. [8] found an allele frequency of 81% in the United Kingdom, and we found a frequency of 64% (not 89% as erroneously stated in the paper) in a large American series [9]. A recent study from Germany [10], not only identified four novel mutations, thus confirming the striking molecular heterogeneity of the disease, but also found a slightly lower allele frequency for the Arg49Stop mutation (56%) than in the British or American cohort. An even lower allele frequency (32%) was found in Italian [11] and Spanish [12] patients. In contrast, the Arg49Stop mutation has not yet been found in any Japanese patient with McArdle disease. It is of practical importance to establish the relative frequency of different mutations in different ethnic groups because screening for the more common "local" mutations in blood DNA can eliminate the need for a muscle biopsy.

Not only is McArdle disease a relatively common cause of myoglobinuria in humans (second only to CPT II deficiency, according to a retrospective study of 77 adult patients conducted by us in 1990 [13]), but it seems to be also common in other animal species. Myophosphorylase deficiency causing exercise intolerance and myoglobinuria has been described in Charolais cattle [14, 15], and in a Merino sheep flock [16]. Both the bovine and the ovine model of McArdle disease will be useful to study the pathogenesis of myoglobinuria and to experiment gene therapy.

A common, and still acceptable, principle of medical school teaching is to avoid diagnosing two genetic diseases in a single patient because the likelyhood of two rare mutations occurring together in the same individual is remote. However, such "double trouble" becomes much more likely when one of the two mutations is common and, by itself, not very deleterious. This is the case for the nonsense C-to-T mutation at codon 12 of the gene encoding the main muscle isozyme of adenylate deaminase (AMPD1): this AMPD1-mut is present in homozygous form in 2% of the population [17]. While a few of these individuals have exercise-related muscle cramps and pains associated with histochemically and biochemically demonstrable AMPD deficiency [18], most are asymptomatic. In a relatively short time, we have observed two children with unusually early onset of myoglobinuria: one had myophosphorylase deficiency and was homozygous for the "common" Arg-49-Stop mutation; the other had phosphofructokinase (PFK) deficiency and was homozygous for the Arg-39-Pro mutation in the PFK gene [19], but both were also homozygous for the AMPD1-mut [20, 21]. Still another young patient with phosphorylase/ AMPD "double trouble" had severe exercise intolerance, multiple episodes of myoglobinuria, and early onset of weakness [22]. Thus, there may well be a synergistic effect of the two mutations: in all three patients, the *per se* mild AMPD1-mut appeared to worsen the phenotypic manifestations of phosphorylase or PFK deficiency. It is, therefore, important to keep in mind the very real possibility of genetic "double trouble" when confronted with a patient whose known metabolic defect results in an unusually severe clinical presentation.

Disorders of lipid metabolism

As mentioned above, CPT II deficiency appears to be the most common metabolic cause of recurrent myoglobinuria in adults [13]. The human CPT II gene has been assigned to chromosome 1p32 [23], and the cDNA has been cloned and sequenced [24]. Four missense mutations have been associated with the typical myopathic adult presentation, of which the most common among European patients was a Ser113Leu mutation [25]. We have confirmed the prevalence of this mutation in American patients through genomic analysis of blood DNA: this is a useful alternative to muscle biopsy, at least as an initial screening [26].

An increasing number of patients with recurrent myoglobinuria and the telltale precipitating circumstances of CPT II deficiency (prolonged exercise, fasting, or a combination of the two) did not have this enzyme defect, and it soon became apparent that defects of β-oxidation could closely simulate the clinical phenotype of CPT II deficiency, including deficiencies of long-chain acyl-CoA dehydrogenase (LCAD), long-chain 3-hydroxyacyl-CoA dehydrogenase (LCHAD), and short-chain 3-hydroxyacyl-CoA dehydrogenase (SCHAD) [27]. A new enzyme defect in the last step of the β-oxidation spiral, *medium chain 3-ketoacyl-CoA thiolase deficiency*, has been reported in an infant with vomiting, liver dysfunction, metabolic acidosis, and terminal massive myoglobinuria [28].

The recent documentation of a higher degree of complexity of the β-oxidation pathway has provided new biochemical causes of recurrent myoglobinuria. In addition to the four "classical" concerted β-oxidation reactions in the mitochondrial matrix (FAD-dependent dehydrogenation of acyl-CoAs; hydration of 2-enoyl-CoAs; NAD-dependent oxidation of 3-hydroxyacyl-CoAs; and CoA-SH-dependent thiolysis of 3-hydroxyacyl-CoAs), two additional enzymes bound to the inner mitochondrial membrane prepare long-chain fatty acids for β-oxidation in the matrix. These are the very-long-chain acyl-CoA dehydrogenase (VLCAD), which acts on acyl CoAs of 20-22 carbon atoms, and the trifunctional protein (TP), which catalyzes all three terminal β-oxidation reactions [27]. Defects of VLCAD and TP have been reported in both children and adults with recurrent myoglobinuria simulating CPT II deficiency [29-32].

Defects of pyruvate ulization

Myoglobinuria had not been reported in association with defects of pyruvate utilization until 1997, when lipoamide dehydrogenase (LAD) deficiency was described in muscle and fibroblasts from a 37-year-old man, who had repeated episodes of acute muscle weakness, vomiting, lactic acidosis, and myoglobinuria, and died during one such episode [33]. Western blots showed that LAD was markedly decreased in muscle mitochondrial fractions, whereas all other components of the pyruvate dehydrogenase complex were present in normal amounts. Sequencing of the coding region of the LAD cDNA failed to reveal a molecular defect.

Defects of the pentose phosphate pathway

In 1987, Bresolin et al. [34] described a young Italian man who, at the end of a 12-km competitive run, lost consciousness and showed signs of both acute hemolysis (jaundice, hyperbilirubinemia, hemoglobinuria) and rhabdomyolysis (serum CK above 40,000 IU, myoglobinuria). Forearm ischemic exercise was normal, and a muscle biopsy showed normal morphology and normal activities for all glycolytic and lipid metabolism enzymes associated with recurrent myoglobinuria. The patient and his two brothers (who had also experienced episodic pigmenturia) had one of the Mediterranean variants of glucose-6-phosphate dehydrogenase (G6PD) deficiency, which typically causes acute hemolytic crises after ingestion of fava (*Vicia faba*) beans or certain drugs. A similar syndrome was documented in three other patients, all of whom had marked deficiency of G6PD in muscle, ranging from 1.3% to 16.5% of normal [35]. As the pentose phosphate is not a major energy-yielding pathway, it was postulated that the myotoxic action of G6PD was mediated through "oxidant stress" due to impaired reduction of glutathione, which is one of the normal functions of G6PD.

Although this form of exercise-induced myoglobinuria is probably rare and may be more common in patients with the Mediterranean variants of G6PD deficiency, in collaboration with Dr. Shin J. Oh (University of Alabama, Birmingham), we have studied an African American boy with an episode of exercise-triggered myoglobinuria. Thorough biochemical analysis of his muscle biopsy did not reveal any known metabolic cause of myoglobinuria, but he was a carrier of the G6PD deficiency trait and G6PD activity in his muscle extract was very low (Di Mauro, Bruno and Oh, unpublished observation).

Respiratory chain defects

Exercise intolerance is the hallmark of mitochondrial myopathies with respiratory chain defects, but it differs from that of patients with glycogenoses in several clinical nuances: (i) "mitochondrial" patients complain of generalized premature fatigue, often described as feeling "out of breath", though they are not necessarily dyspneic; (ii) cramps of exercising muscles occur very rarely, if at all, and (iii) myoglobinuria seems to be less common than in glycogenoses, though it may have been overlooked until recently. Useful and noninvasive diagnostic tests include ^{31}P-NMR spectroscopy and near-infrared spectroscopy. NMR-spectroscopy allows to measure the phosphocreatine (PCr) /inorganic phosphate (Pi) ratio, which in patients is lower than normal at rest, decreases excessively during even moderate exercise, and returns to baseline more slowly than normal during recovery [36]. Near-infrared spectroscopy measures the deoxygenation of hemoglobin during exercise, which is impaired in patients with respiratory chain defects [37].

A few specific biochemical and molecular defects have been recently associated with exercise intolerance, weakness and, in some cases, with recurrent myoglobinuria. The first is coenzyme Q10 (CoQ10) deficiency, which was first reported in 1989 [38] and has been "rediscovered" in recent years [39, 40]. All five patients described so far share the same clinical features: myopathy with recurrent myoglobinuria and ragged-red fibers (RRF), and central nervous system involvement with seizures or mental retardation. This syndrome is probably underdiagnosed, which would be unfortunate because all patients have shown remarkable improvement with CoQ10 supplementation.

It is also apparent that additional different syndromes due to CoQ10 deficiency are now being recognized. A 4-year-old boy with CoQ10 deficiency in muscle had a combination of cerebellar ataxia, seizures, and weakness, associated with increased blood and CSF lactate and MRI evidence of cerebellar atrophy [41]. His strength improved markedly upon CoQ10 administration. We have seen a similar syndrome in three siblings and in an unrelated child, who also responded to CoQ10 therapy [DiMauro and De Vivo, unpublished observation]. In addition, Rötig et al. [42] have reported a more severe form of CoQ10 deficiency in two siblings presenting in infancy with encephalomyopathy and renal failure, also responsive to CoQ10 supplementation. This clinical heterogeneity is not too surprising because the biosynthetic mechanism of CoQ10 is complex, involving formation of the benzoquinone ring to which a side-chain of isoprenyl residues is then attached. Defects at different steps of CoQ10 biosynthesis probably account for the different presentations, but these remain to be defined.

Complex III deficiency accompanies CoQ10 deficiency and should alert to this possibility. However, exercise intolerance with or without myoglobinuria can also occur in patients with isolated complex III deficiency: recently, there has been a series of patients in whom complex III deficiency in muscle was due to distinct mutations in the mtDNA gene encoding cytochrome b. A first missense mutation (G15615A) was found in a 29-year-old man with exercise intolerance and lactic acidosis since 10 years of age [43]. We found a second missense mutation (G15762A) in a 38-year-old woman with exercise intolerance, slowly progressive proximal weakness, and lactic acidosis [44]. A nonsense mutation (G15242A) was discovered in a woman, who had been described 14 years ago as a case of complex III deficiency with exercise intolerance, progressive weakness, and lactic acidosis [45, 46]. At age 17, this young woman became a "cause célèbre" because she benefited from an ingenious therapy, administration of two electron acceptors (menadione and ascorbate) that "bridged" the biochemical gap in the electron transport chain [47]. We have now found another nonsense mutation (G15059A) in a 27-year-old man with exercise intolerance and myalgia since age 10, who has recently had an episode of exercise-induced myoglobinuria [12a]. He is also developing proximal weakness. His muscle biopsy showed RRF but no COX-negative fibers. The lack of central nervous system symptoms in this patient militated against the diagnosis of CoQ10 deficiency, which was excluded by the normal CoQ10 concentration in muscle.

The cytochrome b gene thus appears to be another "hotspot" for mutations in mtDNA, which now have to be included in the differential diagnosis of patients with exercise intolerance, with or without myoglobinuria. These four cases share some interesting features, which are unusual for mtDNA mutations and may have delayed our recognition of these molecular defects: (i) exclusive involvement of muscle; (ii) no evidence of maternal inheritance, and (iii) undetectable levels of the mutations in blood from the patients or their maternal relatives. In addition, the presence of RRF in two of the patients and of "pre-RRF" in one, goes against the still prevailing concept that mutations in mtDNA protein-encoding genes are not associated with substantial mitochondrial proliferation. This concept is also contradicted by findings in patients with mutations in mtDNA COX genes (see below).

Recurrent myoglobinuria (four episodes in one year) precipitated by infection, prolonged exercise, or fasting, dominated the clinical picture in a 16-year-old woman with COX deficiency (14% of normal in isolated muscle mitochondria). The muscle biopsy showed numerous RRF, which, in contrast to those of patients with complex III deficiency, were also COX-negative. Family history was negative. Molecular genetic analysis revealed a 15-bp in-frame deletion in the gene encoding COX III, which removes five amino acids from the third hydrophobic domain of this subunit [48]. The molecular genetic basis of COX deficiency was not clarified in an infant girl, who suffered two episodes of myoglobinuria, one at 3 weeks and the other at 3 months, and finally succumbed to "hemolytic uremia syndrome" at 6 months of age [49]. The enzyme defect, however, had been documented only in fibroblasts and was partial (55% of the lowest control value), and the child had had normal blood lactate and pyruvate, raising some doubts about the primary nature of the COX deficiency.

Myoglobinuria is an unusual clinical manifestation of tRNA mutations in mtDNA, but one such mutation (A606G in the $tRNA^{Phe}$ gene) was reported in a 30-year-old man, who developed myoglobinuria after strenuous lifting [50]. After recovering from the attack of myoglobinuria, his strength was normal, but he had bilateral neurosensory hearing loss, a very common manifestation of tRNA mutations. Also, family history was suggestive of maternal inheritance: a maternal uncle had seizures and two maternal cousins were mentally retarded. Muscle biopsy showed RRF and COX-negative fibers, and COX activity was decreased in muscle.

Conclusions

The list of the metabolic causes of exercise intolerance and myoglobinuria has considerably expanded in the past few years (Table 1). This is mainly due to a better understanding of β-oxidation and its defects, and to the realization that myoglobinuria can occur in patients with defects of the respiratory chain. Among the defects of the respiratory chain, it is particularly important to consider CoQ10 deficiency (Table 2), both because this entity may be more common than we thought and because it is responsive to replacement therapy.

Table 2. Typical clinical features of muscle CoQ10 deficiency

Pt #	CNS signs	Myoglobinuria	RRF & lipid storage	References
1*	seizures	+	+	
2*	mental retardation	+	+	[44]
3	seizures, psychosis	+	+	[12]
4**	seizures	+	+	[14]
5**	seizures	+	+	

Pt: patient; the asterisks indicate two sets of siblings

Acknowledgements

Supported by NIH grants PO1HD32062 and NS11766 and grants from the Muscular Dystrophy Association. Dr. Andreu is supported by the Fondo de Investigaciones Sanitarias (FIS 1005-95); Dr. Bruno by a fellowship from Telethon-Italia, and Dr. Hadjigeorgiou by the Kykko Monastery Foundation, Cyprus.

References

1. Tein I, DiMauro S, Rowland LP (1992) Myoglobinuria. In: Rowland LP, Di Mauro S (eds) Handbook of Clinical Neurology. Elsevier, Amsterdam, Vol 62, pp 553-593
2. Holt IJ, Harding AE, Morgan Hughes JA (1988) Deletions of muscle mitochondrial DNA in patients with mitochondrial myopathies. Nature 331:717-719
3. Wallace DC, Singh G, Lott MT et al (1988) Mitochondrial DNA mutation associated with Leber's hereditary optic neuropathy. Science 242:1427-1430
4. Haller RG, Lewis SF (1984) Pathophysiology of exercise performance in muscle disease. Med Sci Sports Exerc 16:456-459
5. Kreuder J, Borkhardt A, Repp R et al (1996) Inherited metabolic myopathy and hemolysis due to a mutation in aldolase A. New Engl J Med 334:1100-1104
6. Tsujino S, Shanske S, Sakoda S, Fenichel G, Di Mauro S (1993) The molecular genetic basis of muscle phosphoglycerate mutase (PGAM) deficiency. Am J Hum Genet 52:472-477
7. DiMauro S, Servidei S, Tsujino S (1997) Disorders of carbohydrate metabolism: Glycogen storage diseases. In: Rosenberg RN, Prusiner SB, DiMauro S, Barchi RL (eds) The Molecular and Genetic Basis of Neurological Disease. Boston, Butterworth-Heinemann, pp 1067-1097
8. Bartram C, Edwards R, Clague J, Beynon RJ (1993) McArdle's disease: a nonsense mutation in exon 1 of the muscle glycogen phosphorylase gene explains some but not all cases. Hum Mol Genet 2:1291-1293
9. El-Schahawi M, Tsujino S, Shanske S, DiMauro S (1996) Diagnosis of McArdle's disease by molecular genetic analysis of blood. Neurology 47:579-580
10. Vorgerd M, Kubisch C, Burwinkel B et al (1998) Mutation analysis in myophosphorylase deficiency (McArdle's disease). Ann Neurol 43:326-331
11. Martinuzzi A, Tsujino S, Vergani L et al (1996) Molecular characterization of myophosphorylase deficiency in a group of patients from Northern Italy. J Neurol Sci 137:14-19
12. Andreu AL, Bruno C, Gamez J et al (1998) Molecular genetic analysis of McArdle's disease in Spanish patients. Ann Neurol 45:127-130

12a. Andreu AL, Bruno C, Dunne TC, et al. A nonsense mutation (G 15059 A) in the cytochrome b gene in a patient with exercise intolerance and myoglobinuria. Ann Neurol 1999; 45:127-130

13. Tonin P, Lewis P, Servidei S, DiMauro S (1990) Metabolic causes of myoglobinuria. Ann Neurol 27:181-185
14. Angelos S, Valberg SJ, Smith BP et al (1995) Myophosphorylase deficiency associated with rhabdomyolysis and exercise intolerance in six related Charolais cattle. Muscle and Nerve 18:736-740
15. Tsujino S, Shanske S, Valberg SJ, Cardinet GH, Smith BS, Di Mauro S (1995) Cloning of bovine muscle glycogen phosphorylase cDNA and identification of a mutation in cattle with myophosphorylase deficiency, an animal model for McArdle's disease. Neuromusc Disord 6:19-26
16. Tan P, Allen JG, Wilton SD, Akkari PA, Huxtable CR, Laing NG (1997) A splice-site mutation causing ovine McArdle's disease. Neuromusc Disord 7:336-342
17. Morisaki T, Gross M, Morisaki H, Pongratz D, Zollner N, Holmes EW (1992) Molecular basis of AMP deaminase deficiency in skeletal muscle. Proc Nat Acad Sci USA 89:6457-6461
18. Sabina RL, Holmes EW (1995) Myoadenylate deaminase deficiency. In: Engel AG, Franzini-Armstrong c (eds) myology. New York: McGraw-Hill, New York, vol II, pp 1769-1780
19. Tsujino S, Servidei S, Tonin P, Azan G, DiMauro S (1994) Identification of three novel mutations in non-Ashkenazi Italian patients with muscle phosphofructokinase deficiency. Am J Hum Genet 54:812-819

20. Tsujino S, Shanske S, Carroll JE, Sabina RL, DiMauro S (1995) Double trouble: Combined myophosphorylase and AMP deaminase deficiency in a child homozygous for nonsense mutations at both loci. Neuromusc Disord 5:263-266
21. Bruno C, Minetti C, Shanske S et al (1998) Combined defects of muscle phosphofructokinase and AMP deaminase in a child with myoglobinuria. Neurology 50:296-298
22. Rubio JC, Martin MA, Bautista J, Campos Y, Segura D, Arenas J (1997) Association of genetically proven deficiencies of myophosphorylase and AMP deaminase: a second case of "double trouble". Neuromusc Disord 7:387-389
23. Gellera C, Verderio E, Floridia G et al (1994) Assignment of the human carnitine palmitoyltransferase II gene (CPPT II) to chromosome 1p32. Genomics 24:195-197
24. Finocchiaro G, Taroni F, Rocchi M et al (1991) cDNA cloning. sequence analysis, and chromosomal localization of the gene for human carnitine palmitoyltransferase. Proc Nat Acad Sci USA 88:661-665
25. Taroni F, Verderio E, Dworzak F, Willems PJ, Cavadini P, DiDonato S (1993) Identification of a common mutation in the carnitine palmitoyltransferase II gene in familial recurrent myoglobinuria patients. Nature Genet 4:314-320
26. Kaufmann P, El-Schahawi M, DiMauro S (1997) Carnitine palmitoyltransferase II deficiency: Diagnosis by molecular analysis of blood. Mol Cell Biochem 174:237-239
27. Di Donato S (1997) Diseases associated with defects of beta-oxidation. In: Rosenberg RN, Prusiner SB, DiMauro S, Barchi RL (eds) The Molecular and Genetic Basis of Neurological Disease. Boston, Butterworth-Heinemann, pp 939-956
28. Kamijo T, Indo Y, Souri M et al (1997) Medium chain 3-ketoacyl-coenzyme A thiolase deficiency: A new disorder of mitochondrial fatty acid beta-oxidation. Pediatr Res 42:569-576
29. Schaefer J, Jackson S, Dick DJ, Turnbull DM (1996) Trifunctional enzyme deficiency: Adult presentation of a usually fatal beta-oxidation defect. Ann Neurol 40:597-602
30. Smelt AH, Poorthuis BJ, Onkenhout W et al (1998) Very long chain acyl-coenzyme A dehydrogenase deficiency with adult onset. Ann Neurol 43:540-544
31. Minetti C, Garavaglia B, Bado M et al (1998) Very-long-chain acyl-coenzyme A dehydrogenase deficiency in a child with recurrent myoglobinuria. Neuromusc Disord 8:3-6
32. Miyajima H, Orii KE, Shindo Y et al (1998) Mitochondrial trifunctional protein deficiency associated with recurrent myoglobinuria in adolescence. Neurology 49:833-837
33. Elpeleg ON, Saada AB, Shaag A et al (1997) Lipoamide dehydrogenase deficiency: A new cause for recurrent myoglobinuria. Muscle and Nerve 20:238-240
34. Bresolin N, Bet L, Moggio M et al (1987) Muscle G6PD deficiency. Lancet I:212-213
35. Bresolin N, Bet L, Moggio M et al (1989) Muscle glucose-6-phosphate dehydrogenase deficiency. J Neurol 236:193-198
36. Argov Z, Bank WJ (1991) Phosphorus magnetic resonance spectroscopy (31P MRS) in neuromuscular disorders. Ann Neurol 30:90-97
37. Bank W, Park J, Lech G, Chance B (1998) Near-infrared spectroscopy in the diagnosis of mitochondrial disorders. BioFactors 7:243-245
38. Ogasahara S, Engel AG, Frens D, Mack D (1989) Muscle coenzyme Q deficiency in familial mitochondrial encephalomyopathy. Proc Nat Acad Sci USA 86:2379-2382
39. Servidei S, Spinazzola A, Crociani P et al (1996) Replacement therapy is effective in familial mitochondrial encephalomyopathy with muscle CoQ10 deficiency. Neurology 46:A420
40. Sobreira C, Hirano M, Shanske S et al (1997) Mitochondrial encephalomyopathy with coenzyme Q10 deficiency. Neurology 48:1238-1243
41. Boitier E, Degoul F, Desguerre I et al (1998) A case of mitochondrial encephalomyopathy associated with a muscle coenzyme Q10 deficiency. J Neurol Sci 156:41-46

42. Rotig A, Appelkvist E-L, Geromel V et al (1997) CoQ10-responsive mitochondrial encephalomyopathy due to an inborn error of ubiquinone synthesis metabolism. Am J Hum Genet 61(Suppl):A358
43. Dumoulin R, Sagnol I, Ferlin T, Bozon D, Stepien G, Mousson B (1996) A novel gly290asp mitochondrial cytochrome b mutation linked to a complex III deficiency in progressive exercise intolerance. Mol Cell Probes 10:389-391
44. Andreu AL, Krishna S, Bruno C et al (1998) A novel G15762A mutation in the mitochondrial cytochrome b gene. Neurology 50:A24
45. Kennaway NG, Buist NR, Darley Usmar VM et al (1984) Lactic acidosis and mitochondrial myopathy associated with deficiency of several components of complex III of the respiratory chain. Pediatr Res 18:991-999
46. Kennaway NG, Keightley JA, Burton MD, Quan F, Libby BD, Buist NRM (1998) Mitochondrial encephalomyopathy associated with a nonsense mutation in cytochrome b. Mol Genet Metab 63:49
47. Eleff S, Kennaway NG, Buist NR et al (1984) 31P-NMR study of improvement in oxidative phosphorylation by vitamins K3 and C in a patient with a defect in electron transport at complex III in skeletal muscle. Proc Nat Acad Sci USA 81:3529-3533
48. Keightley JA, Hoffbuhr KC, Burton MD et al (1996) A microdeletion in cytochrome c oxidase (COX) subunit III associated with COX deficiency and recurrent myoglobinuria. Nature Genet 12:410-415
49. Saunier P, Chretien D, Wood C et al (1995) Cytochrome c oxidase deficiency presenting as recurrent neonatal myoglobinuria. Neuromusc Disord 5:285-289
50. Chinnery PF, Johnson MA, Taylor RW, Lightowlers RN, Turnbull DM (1997) A novel mitochondrial tRNA phenylalanine mutation presenting with acute rhabdomyolysis. Ann Neurol 41:408-410

Brody's Syndrome: a Disorder of Calcium Uptake by the Sarcoplasmic Reticulum

G. Karpati, D.H. MacLennan

Historical aspects

Lambert and Goldstein [1] published an abstract in 1957, describing a patient with impaired muscle relaxation, mimicking myotonia, in whom electromyographic recording of contracted muscles was "silent". In 1969, Brody [2] published the first complete description of a patient who exhibited impaired muscle relaxation. He was the first to show impaired Ca^{2+}-uptake by sarcoplasmic reticulum isolated from the muscle of his patient.

There were no further reports of this syndrome until 1986 when Karpati et al. [3] presented 4 patients in 2 families suffering from a myopathy with impaired muscle relaxation. These authors recognized the similarity of the clinical phenotype of their patients to Brody's case and proposed that similar cases be designated "Brody's disease". They identified several additional aspects of this disorder: familial incidence with probable autosomal recessive inheritance; absence of Ca^{2+}- ATPase protein in fast - twitch skeletal muscle fibers; and clinical improvement with dantrolene sodium, an inhibitor of the Ca^{2+}- release channel that is widely used as an antidote to malignant hyperthermia reactions [4]. They pronounced Brody's disease to be "a clinically, electromyographically and biochemically distinct metabolic myopathy". Danon et al. [5] confirmed the familial incidence of Brody's syndrome, but also proposed autosomal dominant inheritance for at least one family.

In 1988, Taylor et al. [6] published ^{31}P *in-vivo* nuclear magnetic resonance spectroscopy (NMRS) studies of muscle from a patient suspected of having Brody's disease on the basis of markedly reduced sarcoplasmic reticulum Ca^{2+}- ATPase activity. The NMRS showed reduced high energy phosphate content, increased activity on exercise and post-exercise, but normal ADP recovery rates. Verapamil, a blocker of slow or L-type Ca^{2+} channels (dihydropyridine receptors) located in the sarcolemma and transverse-tubules [7], was beneficial in this patient.

In 1994, Benders et al. [8] published data concerning 10 patients with Brody's syndrome from 7 unrelated families. These authors used immunological and biochemical measurements to show that there was no reduction in Ca^{2+}- ATPase protein content in their patients, but that Ca^{2+}-ATPase activity was reduced, on average, by about 50%. Clinical symptoms in some of their patients were improved with verapamil and also with dantrolene sodium. These authors also demonstrated that cultured Brody myotubes demonstrated prolonged elevation of cytosolic Ca^{2+} after stimulation by acetylcholine.

Table 1. Reports of Brody's disease

Report/Ref	# Of cases	Sex	Age of onset	Family history
Brody [2]	1	F	26	None
Karpati et al. [3]	3	M M M	46 53 Sibs 25	Yes (AR)
Taylor et al. [6]	2	M Sibs F	37 46	Yes
Danon et al. [5]	4	F M Sibs M M	19-54	Yes (AR)
Wevers et al. [21]	1	M	38	Yes
Poels et al. [22]	1	M	38	
Benders et al. [8]	10	M	10-55	Yes (AR)
Odermatt et al. [18]	Idem Karpati et al.; Benders et al.			
Zhang et al. [16]	Idem Karpati et al.; Benders et al.			
Odermatt et al. [19]	Idem Karpati et al.; Benders et al.			

Molecular basis for Brody's disease

Studies beginning in the early 1960s led to the knowledge that skeletal muscle contracts following the sarcoplasmic entry of Ca^{2+} through its release channels located in the sarcoplasmic reticulum [9]. The rise of sarcoplasmic Ca^{2+} concentrations from about 100 nM to 1-10 μM leads to binding of Ca^{2+} to troponin C in the thin filament and induction of the complex changes in the myofibrillar proteins that result in filament sliding and force generation [10]. Skeletal muscle relaxes upon removal of Ca^{2+} by Ca^{2+} pumps located in the same membrane system [9]. These Ca^{2+} pumps are now referred to as sarco(endo)plasmic reticulum Ca^{2+}- ATPases or SERCAs [11].

Three genes encode at least 6 SERCA molecules. SERCA1a and SERCA1b, the developmentally regulated isoforms encoded by the ATP2A1 gene, arise through alternative splicing at the 3' end of the

ATP2A1 transcript [12-14]. SERCA1a accounts for more that 99% of the SERCA isoforms expressed in adult fast-twitch skeletal muscle (type 2 fibers), while SERCA1b predominates in neonatal muscles. SERCA1 isoforms are not expressed to any significant extent in any other tissue but fast skeletal muscle [15]. The human ATP2A1 gene comprises 22 exons spanning about 26 kb of genomic DNA [16] and is located on chromosome 16p.12-p.13 [17]. SERCA2a is highly expressed in slow-twitch skeletal muscle and heart, while the expression of SERCA2b, SERCA3a and SERCA3b is less tissue specific [15].

When the expression patterns of the SERCA genes became evident [15] it was apparent that a mutation in SERCA1 would affect fast-twitch fibers predominantly, while mutation in SERCA2 or SERCA3 would have more generalized effects. On this basis, the reductions in Ca^{2+} pump activity and SERCA protein that had been observed in muscle from Brody patients could be predicted to be result from potential disruptions of the ATP2A1 gene encoding SERCA1.

Odermatt et al. [18,19 and unpublished] have demonstrated mutations in the ATP2A1 gene encoding SERCA1 in 6 unrelated Brody's disease families. In 5 of these families, either homozygous or compound heterozygous mutations would lead to truncation of the SERCA1 protein so that it would be completely inactive. In one family, a homozygous point mutation leads to an expressed, but non-functional form of SERCA1. Thus, the presence of immunostaining for SERCA1 protein in muscle biopsies from Brody patients does not rule out Brody's disease, while the absence of immunostaining is diagnostic of Brody's disease. These results are consistent with the autosomal recessive inheritance of the disease in each of these families.

In 7 other Brody families, no mutations that could cause loss of SERCA1 function were found. This is consistent with studies by Benders et al. [8] in which SERCA1 protein was undiminished in several Brody families, while SERCA1 activity was reduced by up to 50%. Molecules which might modify or stimulate the activity of SERCA1 are possible causes of Brody's disease in non-ATP2A1-linked cases. A prime candidate is sarcolipin (SLN) encoded by the SLN gene [19]. SLN is a 32 amino acid protein which is highly and almost exclusively expressed in fast-twitch muscle. It is associated with SERCA1 and inhibits SERCA1 activity at low Ca^{2+} concentrations by decreasing Ca^{2+} affinity and activates SERCA1 activity at high Ca^{2+} concentrations by increasing Vmax [20]. No mutations have been found in any of the 13 Brody families studied to date [19]. In the families where causal mutations have not been found, further searches must be undertaken for the molecular defect(s) that impairs SERCA1. To date, approximately 22 cases of Brody's syndrome have been reported in the literature (Table 1), but only one-third of these families carry ATP2A1 mutations.

There are no published estimates of the incidence of Brody's disease, partly because the syndrome is difficult to diagnose and partly because of its rarity. The disease is autosomal recessive in the majority of cases, but there are reports of autosomal dominance. It is clearly heterogeneous in terms of molecular etiology. A rough estimate of the incidence of the disease might be 1 in 10,000,000 live births. While this figure indicates that Brody's disease is rare indeed, the incidence of carriers to achieve this incidence would be 1 in 1600, if the disease were invariably autosomal recessive.

Terminology

The designations "Brody's disease" and "Brody's syndrome" have been used for this disorder, without precise definition of the terms. Many but not all features of the clinical phenotype are common to all reported cases of Brody's disease, but, in our experience, Brody's disease is occasionally misdiagnosed by default when major features such as cramping are observed and no more precise etiology can be determined. Even in cases where many features of Brody's disease are present, loss of SERCA1 cannot be demonstrated at the molecular level [8,19]. Thus, current criteria may be descriptive of more than one disease. We propose that Brody's disease be used to describe those cases in which defects in ATP2A1 have been identified, while Brody's syndrome be used for those cases where the precise molecular etiology is unknown.

Clinical features

A fairly stereotyped clinical pattern emerges from the 22 reported cases of Brody's disease that have been described to date [2, 3, 5, 6, 8, 16, 18, 19, 21-23]. The majority of cases suggest an autosomal recessive inheritance but 1 family is reported to have autosomal dominant inheritance [5]. There is a heavy male preponderance. The presenting symptoms are a life-long difficulty in performing sustained, strenuous muscular activity such as running upstairs because the muscles stiffen during exercise and, temporarily, cannot be used. The exercise-induced delay in muscle relaxation involves the legs, arms and eyelids and may be worse in cold weather. Clinical evaluation demonstrates normal strength with a single effort, normal sensation and normal tendon reflexes, but progressive difficulty in relaxing muscles during repeated forceful contraction. Stretch reflexes are not affected. Percussion myotonia is absent. To a variable degree, pain develops in the exercised muscle. Muscle relaxation becomes slower as exercise proceeds and the force of contraction declines. However, quantitative measurements of the decline of twitch and tetanic force after exercise have not yet been reported.

Distal limb and craniofacial muscles are the major muscles involved (Fig. 1), although there appears to be involvement of the bladder sphincter in some cases. Exercise-induced impairment of skeletal muscle relaxation mimics myotonia, while muscle stiffening resembles that observed in metabolic myopathies which are caused by diminished energy supply. Muscle stiffness and pain subside within minutes after cessation of exercise. Fixed muscle weakness or wasting is minimal even years after symptoms are first observed. Indeed, many patients appear to be strongly muscled, perhaps as a result of the demand placed upon slow-twitch fibers which must both compensate for and counteract the diminished relaxation rates of fast-twitch fibers.

Exertional rhabdomyolysis was reported in 1 case [22]. In some cases, the symptoms of Brody's syndrome have been improved by the oral administration of either dantrolene sodium (25 mg 3-4 times per day), verapamil (120 mg 4 times per day) or Nifedipine (10 mg 3 times per day).

Laboratory findings

Electromyographic recordings show that the motor unit potential profile is normal and that there is a normal interference pattern during contraction. There is electrical silence in muscle stiffened by exercise (without volitional contraction). No myotonic or pseudomyotonic discharges are present; hence the term "electrically silent myotonia" [1].

Serum creatine kinase activity is either normal or elevated no more than 5-fold over normal levels.

During routine ischemic forearm exercise tests, lactate production and potassium release from muscle are excessive [21]. Routine histochemistry on muscle biopsies shows moderate to severe histochemical type 2 fiber atrophy. In some cases, immunostaining of histochemical type 2 muscle fibers with either polyclonal or monoclonal antibodies specific to SERCA1 shows a loss of reactivity. As a control, antibodies specific to the slow-twitch/cardiac isoform (SERCA2) is used to show that type 1 fibers show normal immunostaining (Fig. 2). Karpati et al. [3] used an antibody that cross-reacted with both SERCA1 and SERCA2 to show the differential loss of SERCA molecules from fast-twitch fibers.

In the single case that has been investigated [3], the *in vitro* caffeine-halothane contracture tests showed an abnormal response, suggesting susceptibility to malignant hyperthermia. No such reactions have actually been reported for confirmed Brody's disease patients, in spite of the repeated administration of anesthetics for clinical procedures.

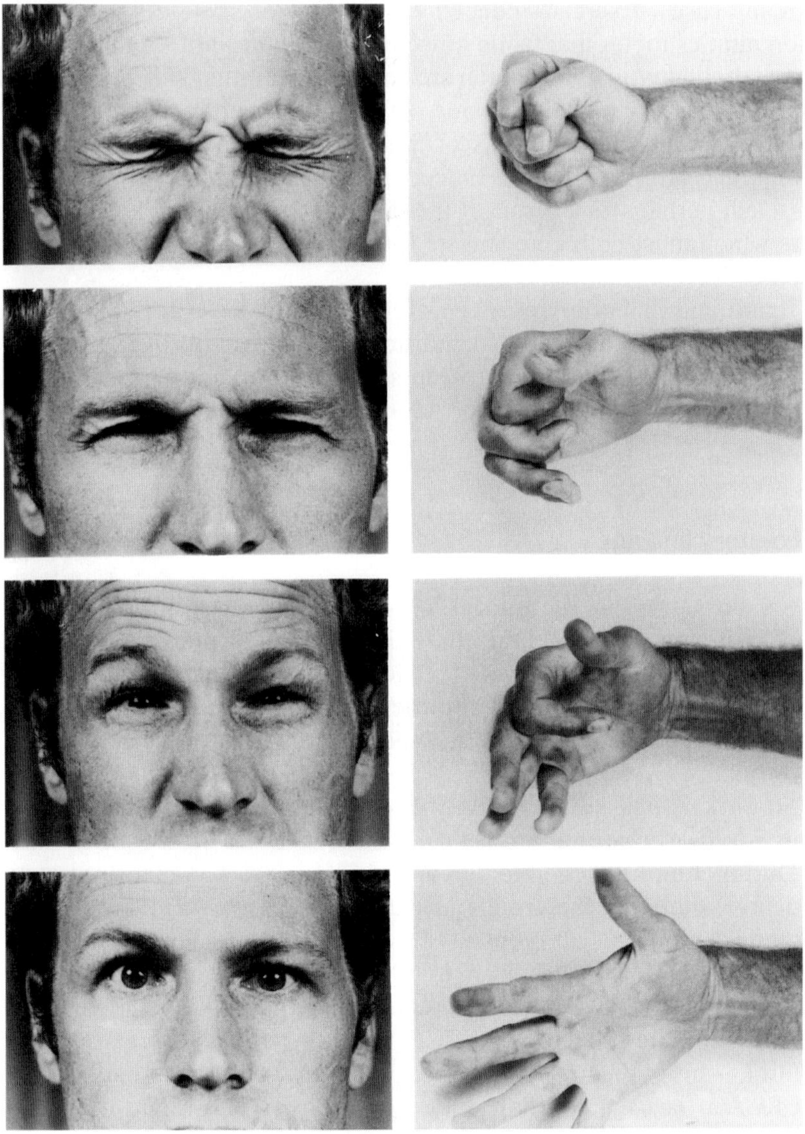

Fig. 1. On the left, the patient tries to open the eyelids after maintaining tight closure for a few seconds. There is an approximately 10 second delay in fully opening the eyelids. In the process, note the recruitment of the frontalis muscles as evidenced by furrowing of the forehead. On the right there is a marked delay in opening a tightly clenched fist, which is highly reminiscent of myotonia. Brody patient with absent SERCA1 and point mutations in the SERCA1 gene producing compound heterozygosity and an unstable truncated protein

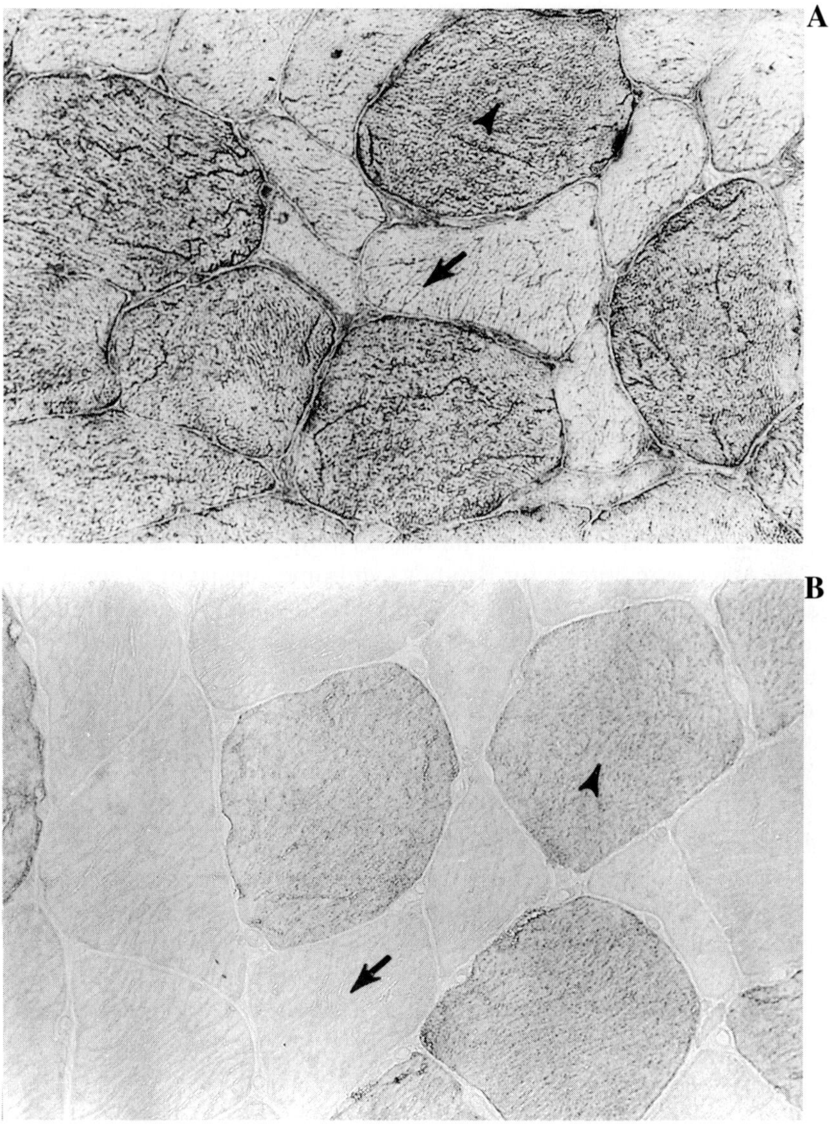

Fig. 2. Immunoreactive SERCA1 and SERCA2 (arrow and arrowhead, respectively) shown by a polyclonal antibody and immunoperoxidase. In panel A, a case of nonspecific type 2 atrophy still shows good SERCA1 protein in the type 2 fibers. In Brody muscle (panel B), the type 2 fibers are devoid of immunoreactive SERCA1, while the type 1 fibers show strong SERCA2 activity. X520

Differential diagnosis

Differential diagnosis must rule out two major disease categories: 1) myotonic disorders such as myotonic dystrophy, myotonia congenita, myotonia fluctuans, paramyotonia or myotonic myopathy, and 2) metabolic myopathies due to either impaired glycogen utilization (for example McArdle's disease) or mitochondrial myopathies due to impaired oxidative phosphorylation. Myotonic disorders can be distinguished by typical electromyographic features while myotonic discharges are absent in Brody's syndrome. Mutation analysis will also distinguish between those genetic defects that cause myotonia and those that cause Brody's disease. McArdle's disease or other glycogen utilization disorders can be identified by biochemical demonstration of the failure of lactate production on exercise, by histochemical or biochemical demonstration of the absence of specific enzymatic activities in muscle biopsies or in muscle homogenates, by the measurement of typical changes in *in vivo* ^{31}P MRS, and by the demonstration of mutations in the gene encoding muscle glycogen phosphorylase.

Mitochondrial myopathies caused by loss of key proteins in the mitochondrial electron transfer chain or oxidative phosphorylation complex can be identified by appropriate histochemistry, by *in vivo* ^{31}P MRS and by the demonstration of causal mutations in mitochondrial DNA.

The diagnostic criteria for Brody's syndrome consist of a typical clinical phenotype, electrical silence in contracted muscle after exercise, reduced uptake of Ca^{2+} into purified microsomal preparations from skeletal muscle biopsies and impaired Ca^{2+}- ATPase activity in muscle *in vitro*. If cytochemical or Western blot analysis or mutational analysis show the absence of SERCA1 protein, the case may be termed Brody's disease.

Cellular and molecular pathogenesis

In Brody's syndrome, Ca^{2+}- uptake into the sarcoplasmic reticulum is impaired due to an insufficiency or absence of Ca^{2+} pump activity. This results in an elevation in the cytosolic Ca^{2+} concentration in stimulated muscle fibers, which increases as exercise progresses, impairing muscle relaxation. Processes known to be Ca^{2+}- dependent could then be activated, accounting for additional deleterious effects on muscle fibers that have been demonstrated in Brody muscle. The activation of phosphorylase kinase by Ca^{2+} leads to the activation of glycogenolysis and glycolysis, leading to heightened lactate production [24). The stimulation of Ca^{2+}- dependent proteases such as cathepsin could stimulate protein catabolism and possibly other pathways leading to apoptosis [25], which might be responsible for the atrophy of type 2 fibers. The elevation of cytosine Ca^{2+} increases the risk of fiber necrosis, the probable mechanism by which muscle proteins enter the blood stream in reported cases of myoglobinuria. The overloading of mitochondria with Ca^{2+} could impair oxidative phosphorylation, leading to diminished energy production.

The therapeutic effect of dantrolene sodium may be explained by its ability to inhibit Ca^{2+}- release through the Ca^{2+}- release channel in the sarcoplasmic reticulum. A reduction of Ca^{2+}- release would compensate for impaired reuptake, leading to less deleterious effects on the fiber. The beneficial effect of T-tubular Ca^{2+}- channel blockers, such as verapamil, can also be explained by a similar mechanism, except that they block the activation of the Ca^{2+}- release channel indirectly through their effect on the voltage sensor of the dihydropyridine receptor, a protein located in the transverse tubular membrane which regulates opening of Ca^{2+}- release channels in the sarcoplasmic reticulum [7].

Conclusions

1 – Brody's syndrome is a rare, familial, disease of skeletal muscle characterized by a life-long difficulty in performing sustained, strenuous muscular activity because of exercise-induced skeletal muscle stiffening, pain and impaired relaxation.
2 – In less than half of the cases studied mutations have been found in the ATP2A1 gene encoding SERCA1. These mutations lead to truncation and absence of SERCA1 protein in fast-twitch skeletal muscle or to expression of SERCA1 with diminished Ca^{2+} pump activity. In the remainder of the cases, further study will be required to identify the precise molecular defect that impairs SERCA1 activity.
3 – The symptoms and signs of Brody's disease and Brody's syndrome can be mitigated by dantrolene sodium and/or verapamil or nifedipine.

References

1. Lambert EH, Goldstein NP (1957) Unusual form of myotonia. Physiologist 1:51 (abstract)
2. Brody IA (1969) Muscle contracture induced by exercise. A syndrome attributable to decreased relaxing factor. N Engl J Med 281:187-192
3. Karpati G, Charuk J, Carpenter S, Jablecki C, Holland P (1986) Myopathy caused by a deficiency of Ca^{2+}- adenosine triphosphatase in sarcoplasmic reticulum (Brody's disease). Ann Neurol 20:38-49
4. Britt BA (1984) Dantrolene. Can Anaesth Soc J 31:61-75
5. Danon MJ, Karpati G, Charuk J, Holland P (1988) Sarcoplasmic reticulum adenosine triphosphatase deficiency with probable autosomal dominant inheritance. Neurology 38:812-815
6. Taylor DJ, Brosnan MJ, Arnold DL et al (1988) Ca^{2+}- ATPase deficiency in a patient with an exertional muscle pain syndrome. J Neurol Neursurg Psychiatry 51:1425-1433
7. Hui CS, Milton RL, Eisenberg RS (1984) Charge movement in skeletal muscle fibers paralysed by the calcium-entry blocker D600. Proc Natl Acad Sci USA 81:2582-2585
8. Benders AA, Veerkamp JH, Oosterhof A et al (1994) Ca^{2+} homeostasis in Brody's disease. A study in skeletal muscle and cultured muscle cells and the effects of dantrolene and verapamil. J Clin Invest 94:741-748

9. Ebashi S, Endo M, Otsuki I (1969) Control of muscle contraction. Q Rev Biophys 2:351-384
10. Zot AS, Potter JD (1987) Structural aspects of troponin-tropomyosin regulation of skeletal muscle contraction. Annu Rev Biophys Chem 16:535-559
11. Burk SE, Lytton J, MacLennan DH, Shull GE (1989) cDNA cloning, functional expression, and mRNA tissue distribution of a third organellar Ca^{2+} pump. J Biol Chem 264:18561-18568
12. Brandl CJ, deLeon S, Martin DR, MacLennan DH (1987) Adult forms of the Ca^{2+}ATPase of sarcoplasmic reticulum. Expression in developing skeletal muscle. J Biol Chem 262:3768-3774
13. Lytton J, Zarain-Herzberg A, Periasamy M, MacLennan DH (1989) Molecular cloning of the mammalian smooth muscle sarco(endo)plasmic reticulum Ca^{2+}- ATPase. J Biol Chem 264:7059-7065
14. Poch E, Leach S, Snape S, Cacic T, MacLennan DH, Lytton J (1998) Functional characterization of alternatively spliced human SERCA3 transcripts. Am J Physiol 275(6 Pt 1):C1449-1458
15. Wu KD, Lee WS, Wey J, Bungard D, Lytton J (1995) Localization and quantification of endoplasmic reticulum Ca(2+) - ATPase isoform transcripts. Am J Physiol 269(3 Pt 1):C775-784
16. Zhang Y, Fujii J, Phillips MS et al (1995) Characterization of cDNA and genomic DNA encoding SERCA1, the Ca(2+) - ATPase of human fast-twitch skeletal muscle sarcoplasmic reticulum, and its elimination as a candidate gene for Brody disease. Genomics 30:415-424
17. Callen DF, Baker E, Lane S et al (1991) Regional mapping of the Batten disease locus (CLN3) to human chromosome 16p12. Am J Hum Genet 49:1372-1377
18. Odermatt A, Taschner PE, Khanna VK et al (1996) Mutations in the gene-encoding SERCA1, the fast-twitch skeletal muscle sarcoplasmic reticulum Ca^{2+}ATPase, are associated with Brody disease. Nat Genet 14:191-194
19. Odermatt S, Taschner PE, Scherer SE et al (1997) Characterization of the gene encoding human sarcolipin (SLN), a proteolipid associated with SERCA1: absence of structural mutations in five patients with Brody disease. Genomics 45:541-553
20. Odermatt A, Becker S, Khanna VK et al (1998) Sarcolipin regulates the activity of SERCA1, the fast-twitch skeletal muscle sarcoplasmic reticulum Ca^{2+}- ATPase. J Biol Chem 273:12360-12369
21. Wevers RA, Poels PJ, Joosten EM, Steenbergen GG, Benders AA, Veerkamp JH (1992) Ischemic forearm testing in a patient with Ca(2+) - ATPase deficiency. J Inherit Metab Dis 15:423-425
22. Poels PJ, Wevers RA, Braakhekke JP, Benders AA, Veerkamp JH, Joosten EM (1993) Exertional rhabdomyolysis in a patient with calcium adenosine triphosphatase deficiency. J Neurol Neurosurg Psychiatry 56:823-826
23. van der Knaap MS, Benders AA, Veerkamp GH, Smit LME (1998) Early-onset Brody's disease. XIV[e] Journee Internationales de Pathologie Neuromusculaire de Marseille
24. Brostrom O, Hunkeler FL, Krebseg (1971) The relation of skeletal muscle phosphorylase kinase by Ca^{2+}. J Biol Chem 246:1961-1967
25. Berridge MJ, Bootman MD, Lipp P (1998) Calcium: a life and death signal [news]. Nature 395(6703):645-648

Stiffness on Exercise: a Non Progressive Disorder of Muscle Function (Brody-Karpati's Syndrome)

T. Kuntzer, R.C. Janzer

Muscle contracture or stiffness occurring with exercise is usually encountered in myotonic syndromes or muscle glycogenoses. However, in 1969 Irwin A. Brody described a new disorder of muscle function characterised by impaired muscle relaxation aggravated by exercise [1]. In this comprehensive paper with report of a single patient, he demonstrated that the most important biochemical finding was a marked decrease in the ability of the isolated sarcoplasmic reticulum to accumulate calcium ions. Reduced uptake of Ca^{2+} in isolated sarcoplasmic reticulum was suggested as the cause of the impaired muscle relaxation. In a series of collaborative studies, George Karpati confirmed that in this disorder a marked reduction of immunoreactive Ca^{2+} ATPase of the sarcoplasmic reticulum (SERCA) of fast - twitch muscle fibers (SERCA1) could be demonstrated [2, 3]. Mutations in the ATP2A1 gene (on chromosome 16q12), encoding SERCA1, have been associated with autosomal recessive inheritance of Brody's syndrome in 3 families, but not with autosomal dominant inheritance of the disease [4]. A search for mutations in the gene encoding sarcolipin, a SERCA1 - associated protein, did not reveal any alterations in 5 families [5].

Since Brody-Karpati's syndrome is relatively rare and genetically heterogeneous, we report here 2 members of a family with this disorder. Unlike other reports, a normal amount of immunoreactive SERCA1 can be demonstrated. In this family the pattern of inheritance appears to be autosomal recessive. In one patient, a second inherited disorder of the renal function was also discovered.

Case reports

Patient 1

This patient came to our attention in 1996, at the age of 44. He was referred to us with the question of a possible relationship between a recently investigated hypokalaemic alcalosis and a long lasting muscle stiffness on exercise. He first sought medical attention at the age of 5. At this age he was not able to run as his friends. When he attempted to run, his legs would stiffen. Thereafter he was aware that muscle stiffness occurred during vigorous or sudden rapid movement such as climbing upstairs, running or screwing on. The stiffness was painless and disappeared after several seconds of rest. Despite these symptoms, he was able to carry out daily activities without difficulty and by pacing his movements, he could ride a bicycle and swim. The condition was non progressive and slightly increased by cold water. A single episode of an unexplained hyperthermia which lead to hospitalisation for 48 hours was noticed during hot weather in his twenties, but several episodes of unknown raised temperature were described. He had been given a medical discharge from military service because of back pain and inability to run. Few episodes of carpopedal spasms had been observed, which had lead to the discovery of a Gitelman's syndrome, a variant of Barrter's syndrome. The patient denied weakness and there was no history of pigmenturia. His parents were first cousins and had no neuromuscular symptoms. Of his brother and sister, a brother (patient 2) is affected by a similar disorder of muscle function. Of his 4 siblings, none is affected by similar stiffness on exercise. Neurological examination was normal except for slow relaxation of limb and face muscles (Table 1). Laboratory studies revealed slight increase level of serum CK; nerve conduction studies and needle EMG examination were normal (Table 2). Patient was treated by potassium repletion with potassium supplements, with no change of his symptoms.

Patient 2

This 40-year-old man (brother of patient 1) had normal developmental milestones. His earliest recollection of a disorder with muscle function was inability to run in foot races at the age of 4 - 5 years. He had the same range of activities as his peers, but could not do them as quickly because he found that vigorous or sudden, rapid movements forced him to stop; strenuous activities such as running produced muscle contracture or stiffness within 2 - 4 seconds, that disappeared after 10 to 20 seconds of rest. The condition was non progressive but in cold weather the stiffness occurred more readily. Despite these symptoms, he was able to carry out daily activities without difficulty and by pacing his movements he could swim and ride a bicycle. He denied weakness and there was no history suggesting pigmenturia. Of his 4 siblings, none is affected by similar stiffness on exercise. Neurological examination was normal except for slow relaxation of limb and face muscles (Table 1). Laboratory studies revealed slight increase level of serum CK and no evidence of Gitelman's syndrome; nerve conduction studies and needle EMG examination were normal (Table 2). The patient was not treated by verapamil.

Table 1. Relevant history information, symptoms and signs of patients 1 and 2

	Patient 1	Patient 2
Duration of the symptoms	Lifelong	
Progression of the symptoms	No	
Ability to do sports	By pacing movements, both patients can ride a bicycle and swim	
Cold aggravation of stiffness on exercise	Slightly	
Episodes of hyperthermia	Several	No
Muscle wasting	No	
Squeletal deformities	Pectus excavatus	No
Normal strength of individual muscles	Yes	
Normal deep reflexes	Yes	
Percussion myotonia (tongue, arm, hand, leg muscles)	No	
On repeated exercises (closing eyelids, making fist, bending ankle up and down), muscle contraction is normal, but the relaxation phase is slow and becomes increasingly slow during exercise (within a minute)	Yes, slow relaxation phase in arms, legs and face muscles	

Muscle biopsy findings

The left peroneus muscle was taken from patient 1 under local anaesthesia. After freezing in liquid nitrogen cooled isopentane, serial cryostat sections were cut at 10 µm and stained conventionally with hematoxylime & eosin, PAS, Oil red O, Engel's modification of trichrome stain, and histochemically for myofibrillar ATP-ase at pH 4.2, 4.5 and 9.4, nicotine adenine dinucleotide tetrazolium reductase (NADH – tr), succinate – dehydrogenase (SDH), cytochrome – C-oxydase (COX), myophosphorylase and phosphofructokinase. The Ca^{2+} – dependent ATP-ase (SERCA 1 and SERCA 2) was visualised by using IgG1 monoclonal mouse antibodies against SERCA 1 and 2 (Novocastra Laboraties Ltd, Newcastle, UK, dilution 1: 200, standard ABC technique, incubation overnight 4°C).

The muscle biopsy showed a good preservation of its fascicular structure with no increase in peri— or endomyseal fibrous or adipose tissue. The number of internal nuclei was slightly increased and reached focally 5%. No degenerating or regenerating muscle fibers were observed. The lipid content was normal. The histochemical expression of all oxydative enzymes examined was normal with no focal increase or decrease in activity. A discrete and selective atrophy of type 2 fibers was documented by all histochemical stains. The immunohistochemical expression of SERCA 1 and 2 was normal (FIGURE 1). We have not examined the activity of SERCA 1 nor the expression of SERCA – related proteins.

Table 2. Results of ancillary tests in patients 1 and 2

	Patient 1	Patient 2
Serum CK	2 – 3 fold increase	3 – 4 fold increase
Other abnormal tests	Hypokalaemic alcalosis, renal magnesium wasting and hypocalciuria (Gitelman's syndrome) High cholesterol level	High cholesterol level
ECG	Normal	
Right median, ulnar, tibial and peroneal nerve conduction studies (distal latency, velocity, amplitude, F-wave latency following 10 distal stimulations)	In the range of normal value	
Bilateral sural nerve conduction study	Normal	
MacManis exercise test (normal decrement in area < 36% and in amplitude < 30%)	Decrement in area = 31.4% and in amplitude = 30.1%	Not done
Needle electromyography at rest (masseter, orbiculairs oculi, biceps brachialis, extensor digitorum communis, first dorsal interossei)	Normal insertion activity. No repetitive discharges or other abnormal potentials	
Needle electromyography during relaxation phase (extensor digitorum communis)	Normal	
Needle electromyography on voluntary muscle activation (extensor digitorum communis muscle)	Normal motor unit potential duration, phase and amplitude of 15 potentials. Turns - amplitude analysis of 20 contractions revealed 5 data points outside of the limiting border (right side) of the normal cloud	

Discussion

Brody's (or Brody-Karpati's) disease is a rare, possibly underrecognised, inherited error of muscle function. Clinical diagnosis is based on a lifelong history of difficulty in performing sustained, strenuous muscular activities, usually such as running upstairs, because of the muscles stiffen during exercise and temporarily cannot be used [1, 3]. The diagnosis can be quite difficult, since the signs and symptoms of the patients are rather heterogeneous and nonspecific. Even impaired muscle relaxation can be absent in some cases as in patients 1, 2, 5, and 9 of Benders et al. [6] and in a patient described by Taylor et al. [7]. Therefore, the diagnosis should be considered more commonly in patients with exercise-induced muscle stiffness, cramp and pain syndromes.

The exercise-induced delay in muscle relaxation involves the legs, arms and eyelids, and may be worse in cold weather, or be associated with hyperthermia. In reported cases, there is, in most instances, no progression of the impairment. Clinical examination demonstrates normal strength with a single effort. There is no muscle wasting or tenderness. All reflexes are present and the plantars flexors. There is no abnormality of sensation or gait. Percussion myotonia is absent. In most cases, diagnosis is therefore based on clinical grounds, after exclusion of other more common causes, such as myotonic syndromes or muscle glycogenoses. Needle electromyography is here very helpful in showing normal spontaneous insertion activity, and no other potentials (no myotonic discharges) at rest and during the delayed relaxation. Histological examination of the muscle biopsy reveals mild fibre atrophy, usually of the fast fibres and is helpful to exclude McArdle's disease.

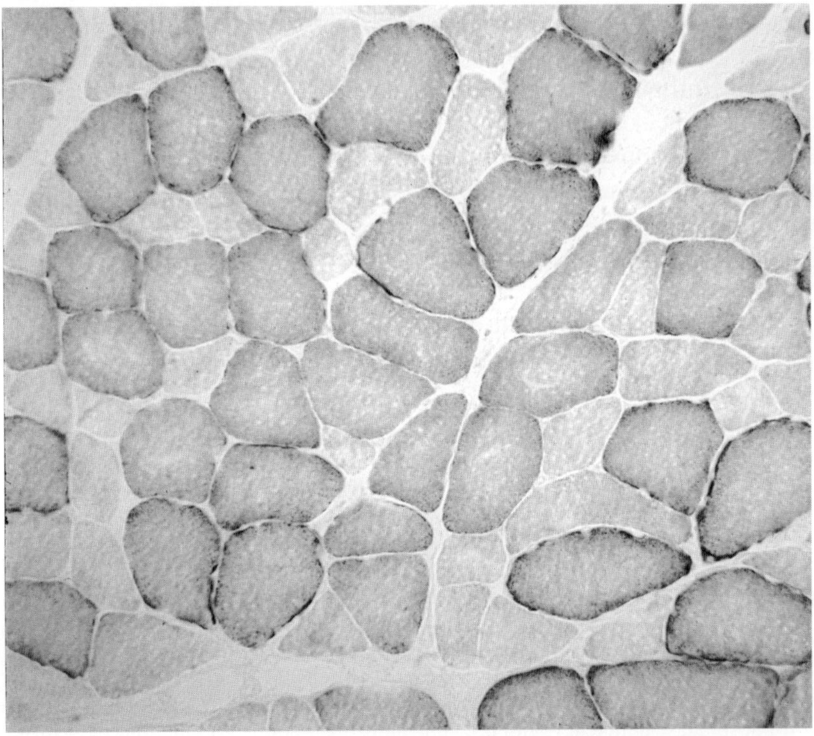

Fig. 1. Left peroneus muscle biopsy from patient 1 stained by IgG1 monoclonal mouse antibodies against SERCA 1 (Novocastra Laboratories Ltd, Newcastle, UK, dilution 1: 200, standard ABC technique, incubation overnight 4°C), showing normal immunostaining of type 2 fibers. Note the good preservation of the fascicular structure. A discrete and selective atrophy of type 2 fibers was documented by all histochemical stains

The diagnosis of Brody-Karpati's syndrome can be confirmed by immuno-histochemical studies on muscle with the demonstration of the absence of SERCA1 protein in type II fibres [2, 3]. Antibodies to SERCA1 and 2 are indeed now available commercially. However, the literature on the content of Ca^{2+}- ATPase protein in Brody patients is apparently conflicting since, as in our patient 1, either a normal content or a 50% reduction in ATPase activity had been reported [6]. Further evaluation of these apparent discrepancies will be necessary to understand why Ca^{2+} ATPase can be present but inactive.

Verapamil or dantrolene sodium can be effective in the syndrome [6, 8] by lowering intracellular Ca^{2+} concentration in the squeletal muscle, but were not taken by our patients.

It is worth noting that patient 1, but not his brother, suffered from Gitelman's syndrome, a variant of Barrter's syndrome (hypokalaemic metabolic alkalosis) with hypomagnesaemia and hypocalciuria. This recessive autosomal syndrome has been recently mapped to chromosome 16, where mutations of the locus encoding for the renal thiazide Na-Cl cotransporter have been demonstrated. Symptoms and signs in affected patients reflect derangements in electrolyte homeostasis and include signs of intravascular volume depletion, seizures, tetany, recurrent cramping with exertion and joint pain with chondrocalcinosis [9,10]. Recurrent back pain and carpopedal spasms reported by our patient were the symptoms related to Gitelman's syndrome, which is treated by potassium repletion.

The study of the reported family, with consanguinity of the 2 parents, will be further investigated by genetical analysis of the ATP2A1 gene and by magnetic resonance spectroscopy.

Acknowledgements

This report would not have been possible without the invaluable cooperation of family members. We are grateful to Dr Cyril Jaques and Christopher Naegeli for referring the patients, to Drs Fulgencio Gomez and François Ochsner for patient care and technical assistance. This work was partly supported by the Association Suisse Romande contre la Myopathie.

References

1. Brody IA (1969) Muscle contracture induced by exercise. A syndrome attributable to decreased relaxing factor. N Eng J Med 281 (4):187-192
2. Danon MJ, Karpati G, Charuk J, Holland P (1988) Sarcoplasmic reticulum adenosine triphosphatase deficiency with probable autosomal dominant inheritance. Neurology 38 (5):812-815
3. Karpati G, Charuk J, Carpenter S, Jablecki C, Holland P (1986) Myopathy caused by a deficiency of Ca^{2+}- adenosine triphosphatase in sarcoplasmic reticulum (Brody's disease). Ann Neurol 20 (1):38-49
4. Odermatt A, Taschner PE, Khanna VK, Busch HF, Karpati G, Jablecki CK, Breuning MH, MacLennan DH (1996) Mutations in the gene-encoding SERCA1, the fast-

twitch skeletal muscle sarcoplasmic reticulum Ca^{2+} ATPase, are associated with Brody disease. Nature Genetics 14 (2):191-194

5. Odermatt A, Taschner PE, Scherer SW, Beatty B, Khanna VK, Cornblath DR, Chaudhry V, Yee WC, Schrank B, Karpati G et al (1997) Characterization of the gene encoding human sarcolipin (SLN), a proteolipid associated with SERCA1: absence of structural mutations in five patients with Brody disease. Genomics 45 (3):541-553
6. Benders AA, Veerkamp JH, Oosterhof A, Jongen PJ, Bindels, RJ, Smit LM, Busch HF, Wevers RA (1994) Ca^{2+} homeostasis in Brody's disease. A study in skeletal muscle and cultured muscle cells and the effects of dantrolene an verapamil. J Clin Invest 94 (2):741-748
7. Taylor DJ, Brosnan MJ, Arnold DL, Bore PJ, Styles P, Walton J, Radda GK (1988) Ca^{2+}-ATPase deficiency in a patient with an exertional muscle pain syndrome. J Neurol Neurosurg Psychiatry 51 (11):1425-1433
8. Poels PJ, Wevers RA, Braakhekke JP, Benders AA, Veerkamp JH, Joosten EM (1993) Exertional rhabdomyolysis in a patient with calcium adenosine triphosphatase deficiency. J Neurol Neurosurg Psychiatry 56 (7):823-826
9. Simon DB, Nelson-Williams C, Bia MJ, Ellison D, Karet, FE, Molina AM, Vaara I, Iwata F, Cushner HM et al (1996) Gitelman's variant of Bartter's syndrome, inherited hypokalaemic alkalosis, is caused by mutations in the thiazide-sensitive Na-Cl cotransporter. Nature Genetics 12 (1):24-30
10. Simon DB, Lifton RP (1998) Ion transporter mutations in Gitelman's and Bartter's syndromes. Cur Opin Nephrol Hypert 7 (1):43-47

Exercise Intolerance in Calpain Deficiency and in α-Sarcoglycanopathy

I. Pénisson-Besnier, I. Richard, F. Dubas, J.S. Beckmann, M. Fardeau

There is evidence that the availability of molecular diagnostic tools has expanded our recognition of the phenotypic expression of muscle disorders. The pseudometabolic expression of dystrophinopathies is now well regognized [1, 2]. Besides dystrophinopathies, other well characterized muscular dystrophies can cause exercice intolerance, as we observed in one patient with limb girdle muscular dystrophy type 2A related to calpain-3 mutations [3] and in two siblings affected by an α-sarcoglycanopathy [4].

The first patient was a 23-year-old woman originating from the Reunion Island. She was referred for muscle symptoms lasting for 3 years. No limitation in physical activities had occurred until she was 20. Three months after a first uneventful delivery, she complained of asthenia, myalgias and lower limb proximal muscles weakness. Clinical examination was mentioned normal, as was a first muscle biopsy. Serum CK level was more than 40 times normal. A polymyositis was suspected and treated with corticosteroids. In absence of improvement, cytotoxic agents were added after a second biopsy which was not more conclusive. Finally, in absence of any benefit, all medications were stopped and the patient had been free of them for ten months at the time we examined her. Her predominant complaints were muscle pain and muscle stiffness induced by everyday activities such as steering her car, carrying her baby. Occasional muscle cramping occurred if contraction was sustained. She did not experience any myoglobinuric episodes. She also complained of progressive proximal weakness affecting lower more than upper limbs. Physical examination demonstrated muscle tenderness. Forearm ischemic exercice induced cramping and a swollen forearm in less than 1 minute. There was a symmetric weakness of latissimus dorsi and serratus magnus. The gait was normal, although there was a slight weakness in the glutei and a CT scan examination revealed symmetric involvement of posterior thigh muscles.

Serum CK level was 25 times normal. Biceps brachialis muscle biopsy showed scattered atrophic fibers, and occasional regenerating fibers.

There was no glycogen or lipid storage. Immunolabelling for dystrophin and sarcoglycans was normal. Because muscle pain was the major symptom, she was investigated for a metabolic myopathy. During exercice on ergometric bicycle, there was an excessive production of lactate at low levels of work (6.8-fold increase of lactic acid concentration for a 65 watt-exercise). NMR spectroscopy was not informative because the exercice was not tolerated. Carnitine palmitoyl transferase assay, fatty-acid oxidation in lymphocytes were normal and no defects were detected in the respiratory chain complexes. The solution of the diagnosis was finally given by the examination of her 26-year-old brother who was not initially known by the proband as being also affected by a muscular disease. His first symptoms occurred at age 18, with difficulty in running. He never complained of muscle pain, even at the early stage of his disease. He presented a typical pattern of limb girdle muscular dystrophy type 2A (LGMD2A) [5] in an advanced functional stage. Examination showed scapular winging and predominating pelvic and posterior thigh muscle weakness whereas quadriceps were less affected. Muscle CT-scan is shown in Fig. 1. CK level was 11 times normal. Brachio-radialis muscle biopsy revealed a moderate variation in muscle fiber diameter and few regenerating fibers. Metabolic studies were not performed because diagnosis of LGMD was made by the initial clinical examination.

Considering the geographic origin of the patients, they were investigated for the presence of mutations in the calpain gene. Both patients were compound heterozygotes carrying the same two mutations: a missense mutation (S744G) and a null mutation (946-1 AG → AA) affecting a splice site.

The following patients were 2 sisters. Clinical onset was around 14 years of age in both of them. They complained of intolerance exercice because of severe myalgias affecting limb muscles and even posterior cervical muscles in one of them. Muscle pain was induced by sport activities in the elder, by running, walking or making repetitives movements in the younger. A painful muscle stiffness remained for several hours after the end of exercice. None of them reported myoglobinuric episodes. They also complained of fatigability when walking, but muscle weakness only became obvious about age 20. Both of them had a mild pelvic girdle weakness whereas thigh muscles were more preserved. Calves muscles were not hypertrophic. A symmetric weakness of scapular muscles was noted in the two sibs, with scapular winging in one case. CK levels were more than 10 times normal in the two sisters. Muscle biopsy showed variation in muscle fiber diameter, increased number of internal nuclei and occasional regenerating fibers.

As for the previous patient, the quality of their complaints prompted us to perform metabolic investigations. Exercise on ergometric bicycle showed an excessive production of lactate at a low level of work in one of them, it was normal in the other. NMR spectroscopy was normal in

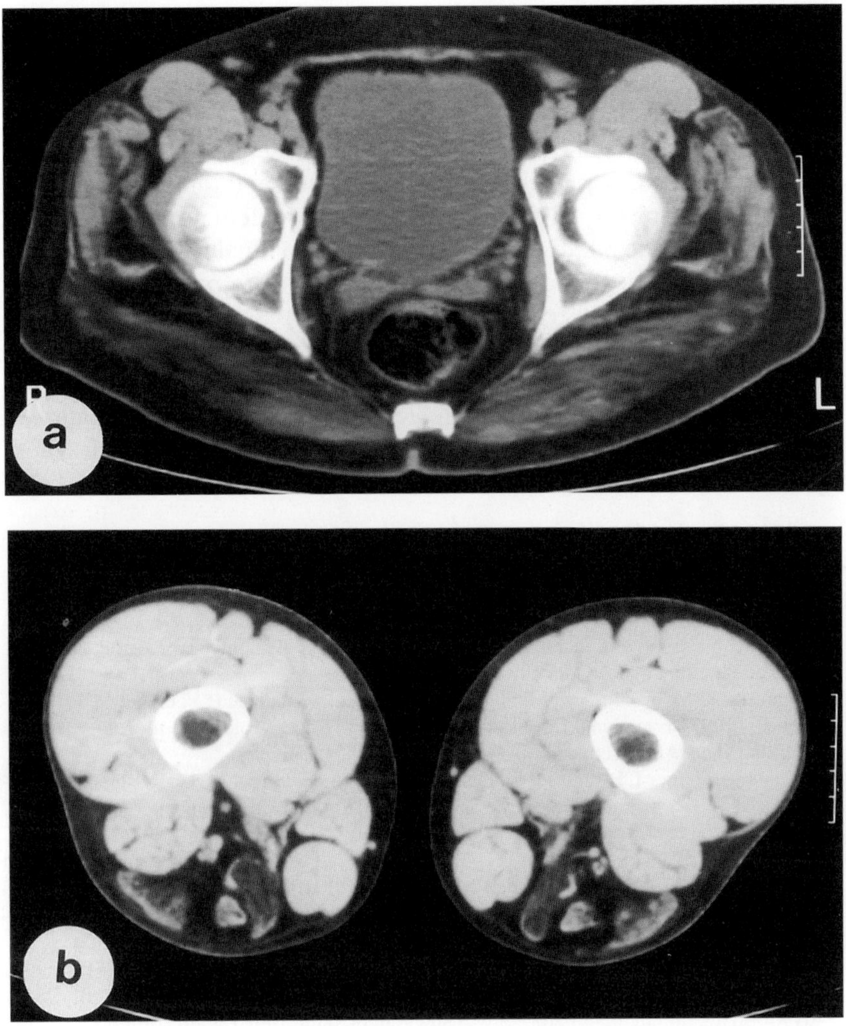

Fig. 1. CT-scan of pelvic (a) and thigh (b) muscles in the patient affected by limb-girdle muscular dystrophy type 2A. Note the selective involvement of posterior thigh muscles

both. Serum carnitine level and glycolysis enzymes in red blood cells were normal.

The diagnosis could be reappraised later on with the availability of anti-sarcoglycans antibodies. The immunohistochemical study showed a normal immunostaining for the anti-dystrophin antibodies and a decreased labelling for the anti α-sarcoglycan antibody. Molecular genetic studies identified a missense mutation in the exon 6 of the adhalin gene.

The metabolic studies performed on these 3 patients, do not allow us to conclude on a type of metabolic dysregulation. An excess of lactate production was inconstantly found and NMR spectroscopy was normal in two of them. Indeed these studies were principally useful to exclude definite metabolic myopathies. We failed to establish any correlation between the pseudo-metabolic phenotype and the type of mutations found either in the adhalin gene [6] or in the calpain gene [7]. Of course further studies are needed, because of the too small number of observations. Furthermore, a completely different aspect of the problem might possibly be considered. It is the putative role of variable individual psychological tolerance to a same degree of weakness. In these three patients, there was an evident contrast between the mild severity of muscle weakness and the severe disability due to muscle pain. Whatever their mechanism may be, the main clinical interest of the cases here reported is to focus on the necessity to search for such protein defects in those patients who complain of exercice intolerance with no or mild muscle weakness, and in whom any evident metabolic dysfunction can be identified.

References

1. Doriguzzi C, Palmucci L, Mongini T, Chiado-Piat, Restano G, Ferrone M (1993) Exercise intolerance and recurrent myoglobinuria as the only expression of Xp2.1 Becker type muscular dystrophy. J Neurol 240:269-271
2. Figarella-Branger D, Baeta Machado AM, Putzu GA, Malzac P, Voelckel MA, Pellissier JF (1997) Exertional rhabdomyolysis and exercice intolerance revealing dystrophinopathies. Acta Neuropathol 94 (1):48-53
3. Pénisson-Besnier I, Richard I, Dubas F, Beckmann JS, Fardeau M (1998) Pseudometabolic expression and phenotypic variability of calpain deficiency in two siblings. Muscle & Nerve 21:1078-1080
4. Fardeau M, Tome FMS, Eymard B, Laforet P, Lombes A, Penisson-Besnier I, Richard I, Leturcq F, Beckmann JS, Kaplan JC (1997) Sur l'expression "pseudo-métabolique" des dystrophies musculaires. In: Livre Jubilaire dédié au Pr Gian Carlo Guazzi, Université de Sienne, pp 279-286
5. Fardeau M, Hillaire D, Mignard C, Feingold N, Feingold J, Mignard D, de Ubeda B, Collin H, Tomé FMS, Richard I, Beckmann J (1996) Juvenile limb-girdle muscular dystrophy. Clinical, histopathological and genetic data from a small community living in the Reunion Island. Brain 119:295-308
6. Piccolo F, Roberds SL, Jeanpierre M, Leturcq F, Azibi K, Beldjord C, Carrie A, Recan D, Chaouch M, Reghis A, E1 Kerch F, Sefiani A, Voit T, Merlini L, Collin H, Eymard B, Beckmann JS, Romero NB, Tome FMS, Fardeau M, Campbell KP, Kaplan JC (1995) Primary adhalinopathy: a common cause of autosomal recessive muscular dystrophy of variable severity. Nature Gen 10:243-245
7. Richard I, Broux O, Allamand V, Fougerousse F, Chainnilkulchai N, Bourg N, Brenguier L, Devaud C, Pasturaud P, Roudaut C, Hillaire D, Passos-Bueno MR, Zatz M, Tischfield JA, Fardeau M, Jackson CE, Cohen D, Beckmann JS (1995) Mutations in the proteolytic enzyme calpain 3 cause limbe-girdle muscular dystrophy type 2A. Cell 81:27-40

Exercise Intolerance and Mitochondriopathies

C. Desnuelle, V. Paquis

Exercise intolerance is the most common complaint in muscle clinic. Patients present with fatigue going worth with exercise and associated with myalgia and/or contracture. It does not include permanent aches or cramps but is sometimes difficult to differentiate from non organic pain syndromes and would possibly remain unexplained after extensive investigations.

Exertional muscle pains may result from either imbalance in muscle energy production and/or utilization during exercise as in metabolic myopathies, ischemia, supraphysiologic exercise or muscle weakness and reduced Vo_2 max due to the loss of functional muscle mass as in muscle dystrophies. In metabolic muscle diseases, isometric exercise is impaired by disorders of anaerobic glycolysis, dynamic exercise is limited by disorders of oxidative metabolism and high-intensity exercise is affected by disorders of mitochondrial pyruvate oxidation, of mitochondrial electron transport chain and of glycolytic pathways when the availability of carbohydrate for oxidation is limited.

In this review we focus on current and personal informations about problems related to the main categories of mitochondriopathies associated with exercise intolerance. We did not included some unusual cases observed by us, for example pyruvate decarboxylase deficiency, glutaric aciduria type II, lipoamide dehydrogenase deficiency, and aconitase deficiency. Therefore, we focus on the two following disorders:

- carnitine palmitoyl transferase (CPT deficiency);
- defects of the mitochondrial respiratory chain.

Carnitine palmitoyl transferase deficiency

CPT deficiency [1] is an example of exercise intolerance in young people. The first attacks began during the first or second decade of life. Sometimes they appear as late as the fifth decade. The majority of patients are male. Female sex is only concerned in 20% of cases.

The symptoms are intermittent. Typically they consist of attacks of muscle pain, cramps, stiffness or tenderness or weakness. Myoglobinuria

[1-3] often is the first symptom [1]. In severe cases the weakness is marked with pigmenturia during several days. The rhabdomyolysis may lead to acute renal failure and to respiratory insufficiency [4].

The attacks of CPT deficiency are most commonly precipitated by heavy exercise, exposure to cold [5], swimming in a cold river [6], fasting, even emotional stress. Cramps, myalgia, muscle weakness develop during prolonged exertion, sometimes several hours after the exercise.

Any muscle group could be involved, for example jaw muscles during prolonged chewing, fingers muscles during prolonged writing. Sometimes the muscle cramps are generalized. The rest do not abort the attack.

Beside these primary forms of CPT deficiency, sometimes apparent causes are noted: pregnancy, viral infection, general anesthesia, surgery, ibuprofen, and diazepam.

There is no therapy other than preventive: high carbohydrate, low fat diet, frequent meals.

Serum creatine kinase is normal or slightly increased at rest, abnormally increased with the prolonged exercise test [7].

The muscle biopsy is usually normal, although an increased lipid deposition has been reported in type I fibers with unspecific myopathic changes. The finding of a normal muscle biopsy does not rule out CPT deficiency. Obviously there may be secondary changes due to the necrosis associated with myoglobinuria.

There is no abnormality in forearm exercise testing, since the forearm test addresses glycogen metabolism rather than lipid metabolism. During the incremental exercise test, the oxygen cost in terms of the amount of oxygen consumed per unit work performed is higher than normal. Because the patient with CPT deficiency has difficulty metabolizing fatty acids, the respiratory exchange ratio (RER, respiratory quotient) is higher or more characteristic of carbohydrate metabolism than fatty acid metabolism. The hyperventilation threshold or anaerobic threshold is higher than the normal value. With the prolonged exercise test, there is an abnormal increase in serum CK for the level of exercise undertaken. It should be stressed that prolonged exercise testing in the patient with CPT deficiency runs the risk of producing myoglobinuria. Fasting is also a provocative test in this disorder. An increase in CK and serum myoglobin may be noted during the fast. There is also a deterioration in the work performance of patients after fasting. The enzyme defect is confirmed in all patients; CPT activity is undetectable in muscle, leukocytes, platelets and cultured fibroblasts [8]. The defect of CPT, irrespective of which of the two isoenzymes CPT I or II is more affected, would result in impaired transport of long-chain fatty acids across the mitochondrial inner membrane and, therefore, impaired oxidation. This metabolic defect can explain many of the clinical features of CPT deficiency. Long-chain fatty acids are the main source of energy for muscle at rest and during moderate exercise of long duration, while muscle appears to depend on muscle glycogen for heavy exercise. Glycogen metabolism is normal in these patients and so is their capacity for strenuous exercise of short duration. With prolonged activity, however, an energy failure develops in CPT deficiency patients as the dependence on long-chain fatty acids increases. The situation is worsened by fasting, because glucose supply from liver glycogen is probably decreased and ketone body formation is inadequate. Whether this energy failure translates into decreased ATP concentration in muscle is not known, and the pathogenesis of myoglobinuria is uncertain. CPT deficiency is an autosomal recessive disorder. The gene encoding CPT is located on chromosome 1 p 32 [9, 10].

Exercise intolerance and defects of the mitochondrial respiratory chain

Mitochondrial cytopathies are disorders of primary oxidative mitochondrial metabolism involving the electron transfer process through the enzyme complex I to IV. They result in multisystem diseases affecting tissues that require intact oxidative metabolism in which muscle involvement is a common but not invariable characteristic.

We will first consider our experience, then summarize the literature.

Exercise intolerance and mitochondriopathy in our experience

We report our data considering first, patients seen in our out-patient clinic complaining of exercise intolerance, and on second time information given when sample were referred to our laboratory for mitochondrial DNA (mtDNA) analysis.

Exercise intolerance and mitochondriopathy

During the last five years, 48 patients aged from 22 to 74 (female: 31, male: 17) have been investigated for the main complaint of exertional muscle pain. Clinical examination was normal in 43, in 5 ptosis, upper limb girdle weakness, calf hypertrophy, brisk tendon reflexes or myotonia were present. Ancillary examination revealed elevated CK level (> 2 N) in 6/48, elevated SR in 2/48, serum lactate, pyruvate ratio (L/P) at rest more than 20 in 2/36, electromyographic myopathic pattern in 8/48 and abnormal muscle ^{31}P NMR spectroscopy in 3/3. The muscle biopsy was normal in 21 patients or with no specific changes for type II fibers atrophy or central nuclei in 7. Dystrophic finding were observed in 2, tubular aggregates in 1 and significant inflammatory pattern in 4. Less than 5% or Cox negative fibers or RRF was noted in 9 patients. One had acid maltase deficiency. The diagnosis of mitochondrial myopathy was affirmed only in 3 patients having more than 5% of Cox negative fibers or RRF. Patient

1 was a 25 year-old male with ocular myopathy, normal CK at rest but L/P > 20. Biochemical analysis demonstrated a combined complex 1 and IV defect but no mutation was detected in mt DNA. Patient 2 was a 53 year-old female presenting with limb girdle weakness. At rest CK was elevated but L/P normal. Biochemical analysis of muscle sample showed complex III defect and no mutation was detected in mt DNA. Patient 3 was a 35 year-old male with normal clinical examination. CK and L/P at rest were normal; biochemical analysis of muscle revealed a complex IV defect and molecular study demonstrated an heteroplasmic 3,5 Kb deletion in mt DNA.

Thus, we considered in our experience that only 3 patients over 48 complaining of exercise intolerance had a mitochondrial cytopathy, 1 of them presenting with normal examination.

Mitochondrial cytopathy and exercise intolerance

We considered for the five past years 106 patients (age ranging from 15 to 78, male 51, female 55) because the presence of exercise intolerance among the clinical features documented with the sample over 621 adult patient samples referred to our laboratory for mtDNA analysis. All had RRF or Cox negative fibers considered as significant by an experienced pathologist and 23% had deletion or point mutation of mt DNA. Among them, exercise intolerance was the only clinical feature in 5, others having exercise intolerance included in various phenotype as ocular myopathy [11, 25], weakness of extra-ocular muscles [12], diabetes and deafness [1], episodic encephalopathic manifestation [2], permanent myalgia [13], cardio-myopathy [14], rhabdomyolysis [15] or Leber optic atrophy [16]. Thus, in this population 5/621 patients had isolated exercise intolerance, 101/621 presented with exercise intolerance as an additional trait of heterogeneous phenotypes.

Overview of the literature

Exercise intolerance is not frequently reported to mitochondrial respiratory chain defect.

Complex I deficiency

Complex I (NADH ubiquinone reductase) carries hydrogen from NADH to coenzyme Q. In these cases exercise intolerance is characterized by isolated muscle pain without cramps or myoglobinuria. Symptoms appear in childhood and early adolescence, or later [11, 25]. A myopathy with limb weakness is often associated [12, 17-19]. In some cases episodes of weakness are accompanied by vomiting and headache. These episodes are precipitated not only by exercise but by alcohol or infection. Lactic acidosis is usual, enhanced by exercise. Muscle biopsy showed ragged red fibres and, with electron microscopy, abnormal mitochondria. A partial combined deficiency of complex IV is observed in about half of the patients.

Recently, a family was reported in which 15 members over 24 in 5 generations presented with exercise intolerance combined with proximal weakness and lactic acidemia [20]. Six patients were analyzed with defect in complex I in 1, in complex II in 2 and in complex IV in 2.

Phosphorus nuclear magnetic resonance spectroscopy of muscle during exercise is a good approach [15, 16], showing a low Pc_r/Pi ratio at rest, a rapid decrease of PC_r during exercise, and a delayed recovery at the end of exercise. In spite of lactic acidosis pH is unchanged.

Some cases could improve with riboflavin [19] but are later complicated by chronic muscle weakness.

Complex III deficiency

Exercise intolerance has been described in 3 cases [13, 14]. Symptoms begin in childhood or adolescence. Lactacidaemia is variable, often enhanced by exercise. A fixed weakness may be preceded by years of exercise intolerance. Muscle biopsy shows ragged red fibers [21]. Vitamine K3 (menadion) and vitamin C could improved exercise capacity.

Coenzyme Q10 deficiency

Exercise intolerance with myoglobinuria is associated with seizures, epilepsy cerebellar ataxia [22] in rare cases. Lactic acidaemia, lipid myopathy, mitochondrial abnormalities are present. This encephalomyopathy is probably related to a defect of a specific enzyme involved in the CoQ10 synthesis in brain and muscle mitochondria. Replacement by ubiquinone therapy is effective [23].

Defects of complex IV

Exercise intolerance with lactic acidosis and abnormal pulmonary regulation in exercise related to muscle cytochrome c oxidase deficiency is observed in adults [24]. Muscle biopsy shows lipid myopathy in type I fibers, increased number of mitochondria with or without ragged red fibres and, of course, Cox negative fibers.

Conclusion

Exercise intolerance is a very common complaint. Extensive investigations are often necessary to objective a metabolic defect affecting the oxidative pathway. CPT deficiency has to be considered in young patient and especially when the muscle biopsy shows normal pattern. According with our experience and in reference with the literature, it appears that less than 1% of patients presenting with isolated exercise intolerance have mitochondrial respiratory chain defect. But exercise intolerance is present in more than 15% of adult patients with mitochondriopathy. In this occurrence it is more frequently associated with ocular myopathy, weakness, diabetes and deafness or episodic lactic acidaemia. In some cases the identification of the defect may allow symptomatic therapy.

References

1. Di Mauro S, Melis-Di Mauro PM (1973) Muscle carnitine palmitoyltransferase deficiency and myoglobinuria. Science 182:929
2. Bank WJ, Di Mauro S, Bonilla E et al (1975) A disorder of muscle lipid metabolism and myoglobinuria. Absence of carnitine palmitoyltransferase. N Engl J Med 292:443
3. Herman J, Nadler HL (1981) Recurrent myoglobinuria and muscle carnitine palmitoyltransferase deficiency in fibroblasts from a patient with muscle CPT deficiency. Ann Neurol 10:196
4. Bertorini T, Yeh YY, Trevisan C et al (1980) Carnitine palmitoyltransferase deficiency: Myoglobinuria and respiratory failure. Neurology 30:263
5. Broownel AKW, Severson DL, Thomson CD, Fletcher T (1979) Cold induced rhabdomyolysis in carnitine palmitoyltransferase deficiency. Can J Neurol Sci 6:367
6. Desnuelle C, Pellissier JF, De Barsy T, Serratrice G (1990) Intolerance à l'exercice par déficit en carnitine palmityl transferase. Rev Neurol (Paris) 146:231-234
7. Carroll JE, Brooke MH, DeVivo DC et al (1978) Biochemical and physiological consequences of carnitine palmitoyltransferase deficiency. Muscle Nerve 1:103
8. Pula TP, Max SR, Zielke HR et al (1981) Selective carnitine palmitoyltransferase deficiency in fibroblasts from a patient with muscle CPT deficiency. Ann Neurol 10:196
9. Finocchiaro G, Taroni F, Rocchi M et al (1991) EDNA cloning sequence analysis, and chromosome localization of the gene for human carnitine palmitoyltransferase. Proc Natl Acad Sci USA 88:661-665
10. Taroni F, Verderio E, Dworzak R et al (1993) Identification of a common mutation in the carnitine palmitoyltransferase gene in familial recurrent myoglobinuria patients. Nature Genet 4:314-319
11. Morgan-Hughes JA, Cooper JM, Holt IJ et al (1990) Mitochondrial myopathies. Clinical defects. Biochem Soc Trans 18:523
12. Watmouth MJ, Bindoff LA, Birch-Machin MA et al (1990) Impaired mitochondrial β-oxidation in a patient with an abnormality of the respiratory chain. J Clin Invest 85:177
13. Dumoulin R, Sagnol I, Ferlin T, Bozon D, Stepien G, Mousson B (1996) A novel gly290asp mitochondrial cytochrome b mutation linked to a complex III deficiency in progressive exercise intolerance. Mol Cell Probes 10:389-391
14. Bouzidi MF, Schagger H, Collombet JM, Carrier H, Flocard F, Quard S, Mousson B, Godinot C (1993) Decreased expression of ubiquinol-cytochrome c reductase subunits in patients exhibiting mitochondrial myopathy with progressive exercise intolerance. Neuromuscul Disord 3:599-604
15. Arnold DL, Taylor DJ, Radda GK (1985) Investigation of human mitochondrial myopathies by phosphorus magnetic resonance spectroscopy. Ann Neurol 18:189
16. Argov Z, De Stefano N, Taivassalo T, Chen J, Karpati G, Arnold DL (1997) Abnormal oxidative metabolism in exercise intolerance of undetermined origin. Neuromuscul Disord 7: 99-104
17. Roodhooft AM, Van Acker KJ, Martin JJ et al (1986) Benign mitochondrial myopathy with deficiency of NADH-CoQ reductase and cytochrome c oxidase. Neuropediatrics 17:221
18. Koga Y, Nonaka I, Kobayashi M et al (1988) Findings in muscle in complex I (NADH coenzyme Q reductase) deficiency. Ann Neurol 24:749

19. Arts WFM, Scholte HR, Bogaard JM et al (1983) NADH-CoQ reductase deficiency myopathy: Successful treatment with riboflavin. Lancet 2:581
20. Heiman-Patterson TD, Argov Z, Chavin JM, Kalman B, Alder H, Di Mauro S, Bank W, Tahmoush AJ (1997) Biochemical and genetic studies in a family with mitochondrial myopathy. Muscle Nerve 20:1219-1224
21. Schapira AHV, Cooper JM, Morgan-Hughes JA et al (1990) Mitochondrial myopathy with a defect of mitochondrial protein transport. N Engl J Med 323:37
22. Ogasahara S, Engel AG, Frens D, Mack D (1989) Muscle coenzyme Q deficiency in familial mitochondrial encephalomyopathy. Proc Natl Acad Sci USA 86:2379
23. Servidei S, Spinazzola A, Crociani P et al (1996) Replacement therapy is effective in familial mitochondrial encephalomyopathy with muscle CoQ10 deficiency. Neurology 46:A420
24. Haller RG, Lewis SF, Eastabrook RW et al (1989) Exercise intolerance, lactic acidosis and abnormal pulmonary regulation in exercise associated with adult skeletal muscle cytochrome c oxidase deficiency. J Clin Invest 84:155
25. Schapira AHV, Cooper JM, Morgan-Hughes JA et al (1988) Molecular basis of mitochondrial myopathies: Polypeptide analysis in complex I deficiency. Lancet 1:150

Dystrophinopathies in Exercise Intolerance

B.A. KAKULAS

Exercise intolerance in the dystrophinopathies is characterised by a group of symptoms which reduce independent mobility. These symptoms are muscular weakness, excessive fatigability, pain and stiffness. Exercise intolerance is a major consideration within the symptomatology of Duchenne muscular dystrophy (DMD) and to a lesser extent, the other types of dystrophinopathy. As a primary muscle disease, weakness and fatigue are paramount in DMD while stiffness and muscle cramps are of lesser importance. Exertional pain is usually not encountered in DMD. Symptoms are similar in Becker muscular dystrophy (BMD) and other members of the dystrophinopathy group, but are less severe.

The dystrophinopathies

Dystrophin is a large cytoskeleton protein (427 Kd) related to the spectrin family which underlies and gives strength to the muscle membrane. A dystrophinopathy is a genetic disorder which results from an abnormality of dystrophin (Table 1). This may take the form of total absence of dystrophin as in DMD or of a reduced quantity and/or molecular size as in BMD. Similarly, truncated forms of dystrophin account for a type of hereditary cardiomyopathy. In the same vein, minor abnormalities of dystrophin may simply manifest as myalgia and cramps or account for asymptomatic creatine kinase aemia [1].

Each of the dystrophinopathies is caused by a mutation of the dystrophin gene which is at locus xp21.2 on the X chromosome. Symptoms arise when muscle membrane integrity fails due to the dystrophin abnormality. The most common type of mutation occurring in approximately 65% of DMD cases is a deletion of DNA. Duplications and point mutations accounting for the majority of the remainder of DMD patients. As a broad principle, the mutation in DMD is non-sense or null type and is "out of frame" so that no dystrophin is formed. The severe DMD phenotype is therefore due to total absence of dystrophin. In BMD and the other dystrophinopathies, the mutation is "in frame" or missense type so that a truncated form of dystrophin is produced. In general, the more dystrophin present, the clinically less severe the disorder.

Table 1. The human dystrophinopathies (Xp21.2 myopathies)

- Duchenne muscular dystrophy (DMD)
- XO or translocated females showing DMD phenotype
- Manifesting (female) carriers of DMD
- Becker muscular dystrophy (BMD)
- Atypical BMD
- Myocardiopathy with proximal weakness
- Pure myocardiopathy
- Myalgia and cramps
- Creatine-kinase-aemias

The situation in the manifesting Duchenne female carrier is slightly different. Due to the lyonization of the X chromosome during development, a proportion of the muscle fibres within the musculature of the female carrier will have total absence of dystrophin. The remainder show normal amounts, each nucleus reflecting the mutated gene in the affected X chromosome or the normal X, depending which nucleus is lyonized. Because the muscle fibres lacking dystrophin are prone to early disruption and disappearance as the carrier becomes older, the normal muscle fibres enlarge and compensate to some extent for those which are lost. The carrier becomes symptomatic when a significant proportion of such fibres are lost and the compensating processes are not fully adequate.

To be distinguished from the dystrophinopathies are several autosomal recessive forms of limb girdle muscular dystrophy which are collectively known as the sarcoglycanopathies. These conditions are named according to their molecular genetic abnormality. In common with DMD and BMD, a null mutation results in a severe Duchenne-like phenotype for the sarcoglycanopathies and an "in frame missense" mutation which allows some of the product to be formed, accounts for a less limb girdle severe phenotype (Table 2).

Exercise intolerance

It is evident from the above that exercise intolerance refers to symptoms which result from impaired muscle function and are thus dynamic in nature and are due to reduced force generation. Rapid fatigue is usually interpreted by the patient as either weakness or lack of stamina. Other symptoms resulting in exercise intolerance are cramps, pain and muscle stiffness or rigidity (Table 3).

Table 2. The sarcoglycanopathies

	mw	Designation	Inheritance	Locus
– darcoglycan	35kd	limb girdle	MD 2C	AR 13q12kd
– sarcoglycan (adhalin)	50kd	limb girdle	MD 2D	AR 17q12-q
– sarcoglycan	43kd	limb girdle	MD 2E	AR 4q12
– sarcoglycan	35kd	limb girdle	MD 2F	AR 5q33-q34

Symptoms of exercise intolerance in DMD and BMD

Symptoms generally develop at the age of five or six years in boys with DMD or BMD. This is usually in the form of being unable to keep up with peers in running and with frequent falls and in particular, climbing. Difficulty in climbing steps is so great that it is used as a definition consideration within the Vignos functional scale [2] in the stages of DMD (Table 4).

Table 3. Symptoms of exercise intolerance

- Weakness
- Stiffness and rigidity
- Fatigue
- Pain and cramps
- Respiratory and cardiac failure

Although symptoms of weakness become apparent at the age of five or six years in DMD or BMD, abnormalities of muscle structure and function may be detected objectively at much earlier stages. Generally, but not universal, there is a delay in motor development in infancy such as crawling and walking. Toe walking is an early feature and the gastrocnemius muscles may be larger and firmer than in other boys of the same age. An exaggerated lumbar lordosis may also be apparent. However, these presymptomatic signs may be difficult to detect or absent.

The pattern of muscle involvement is very consistent, being bilaterally symmetrical and involving the girdle and proximal limb muscles most severely, particularly in the lower limbs. A characteristic feature is the selectivity of muscle involvement, the quadriceps, iliopsoas, and glutei being particularly involved in the lower limbs and the latissimus dorsi, serrati, sternocostal head of pectoralis major, biceps, triceps and brachioradialis in the upper limbs.

Weakness of the knee and hip extensors results in the classical Gowers sign when the child attempts to rise from the floor and weakness of the hip extensors leads to the development of an accentuated lumbar lordosis.

At an earlier stage, even before the Gowers sign can be elicited, affected boys may be unable to rise from a sitting position on the floor if asked to do so with the arms folded. The gait subsequently becomes waddling in character owing to the weakness of the hip abductors and there is often a tendency to walk on the toes. Pseudohypertrophy of the calves, which is due largely to the accumulation of adipose and connective tissue, is a distinctive feature in over 90% of cases, the calf muscles having a firm, doughy or woody consistency on palpitation. Other muscle groups, such as the deltoids, wrist extensors, spinati, masseters and the tongue, may also be enlarged in some patients [3].

Table 4. The Vignos scale

- Walks and climbs stairs without assistance
- Walks and climbs stairs with aid of railing
- Walks and climbs stairs slowly with aide of railing (25 sec for 8 standard steps)
- Walks unassisted and rises from chair but cannot climb stairs
- Walks unassisted but cannot rise from chair or climb stairs
- Walks only with assistance or walks independently with long leg braces
- Walks in long leg braces but requires assistance for balance
- Stands in long leg braces but unable to walk even with assistance
- In wheelchair or bed

Contractures result from immobility deriving from weakness and consequent postural deformities. The imbalance between opposing muscles aggravates the contractures, eg the anterior tibial group versus gastrocnemius and soleus or the quadriceps versus the hamstring muscles.

As the disease advances there is progressive atrophy and weakness of the proximal limb and girdle muscles and the differential pattern of muscle involvement becomes progressively less obvious. The deep tendon reflexes are depressed, with the exception of the ankle jerks, which are preserved even in the late stages of the disease [4].

Muscle fatigue or loss of stamina is a symptom common to all myopathies. In this respect, the patient is unable to sustain a particular motor activity for as long as normal individuals. Again, fatigue in DMD and BMD is directly related to loss of muscle fibres within the most affected groups. The advent of cramps seems to be the result of increased irritability of muscle in the early stages of the disorder. This may be related to entry of calcium due to the absence of dystrophin and the consequent abnormal membrane, but scientific data on this aspect is lacking. Muscle stiffness or rigidity may be a feature in the later stages of the disorder reflecting the fat and fibrous tissue replacement of muscle fibres.

Pain is not a feature of the dystrophic process. However, the postural imbalance as a result of differential weakness of muscles may be painful due to pressure or if it results in dislocations.

Cardiac and respiratory symptoms

Cardiac involvement is most frequent in DMD in which fibrosis, particularly of the left ventricular myocardium and conducting system, is invariably found at postmortem. During life, asymptomatic sinus tachycardia, arrhythmias, mitral valve prolapse and other non-specific murmurs are common as are electrocardiographic abnormalities, which are present in most cases and include conduction defects, tall R waves in right precordial leads and deep Q waves in the limb and left precordial leads. Echocardiography shows characteristic changes in the contractility of the free posterior left ventricular wall which increase with the passage of time. Sudden death may occasionally occur as a result of an arrhythmia, but progressive cardiac failure is unusual in DMD.

The heart is affected in DMD because of total absence of dystrophin. In DMD, at autopsy, cardiac muscle fibres are lost and are replaced by dense connective tissue which surrounds islands of surviving myofibres. The changes are most severe in the left ventricular myocardium and conducting system. Cardiomyopathy in patients with BMD may be a prominent feature, especially in some families and leads to cardiac failure as the main cause of death. There are also reports of isolated cases or families with cardiomyopathy as the only or major manifestation of a dystrophinopathy. A number of these patients have had deletions involving exon 1 or the muscle promoter region of the dystrophin gene. The partial absence of dystrophin is responsible for the cardiac changes in the other dystrophinopathies.

Dyspnoea is not usually a feature of either DMD or BMD because of the muscular weakness which rarely reaches the point of increased demand for oxygen. However, respiratory failure due to involvement of the intercostals and diaphragm is the common terminal event in DMD.

Correlation of symptoms with muscle lesions in DMD

The absence of dystrophin in DMD predisposes to destruction of muscle fibres presumed to be due to the entry of calcium as a result of tears in the muscle membrane. The role of dystrophin is presumed to be primarily mechanical, providing strength to the muscle membrane. Contractions, especially those of the eccentric type, create tension on the membrane which then, without the normal cytoskeleton support which would be provided by dystrophin, undergoes disruption. The entry of calcium activates proteases and causes destruction of the contractile machinery. Pathologically, this biochemical damage manifests as necrosis and sets in train a well established sequence of events beginning with myophagia by macrophages and ending with incomplete attempts at regeneration [5]. This is a cyclical process which eventually, due to the exhaustion of regeneration, leads to outfall of muscle fibres which are replaced by fat and fibrous tissue. Remaining muscle fibres undergo hypertrophy or atrophy when disrupted at some point in their length. The outfall of muscle fibres as a result of this process is mainly responsible for muscle weakness and associated symptoms of exercise intolerance in DMD and BMD. One of the remarkable and not fully explained features of DMD and BMD is the selective involvement of the large muscles of the body within the limb girdles. Especially affected are the adductors of the flexors and adductors of the hips, the gluteal muscles, hamstrings and iliotibial band. In the upper limbs, the periscapular muscles and pectorals are more affected. The distal muscles and those of the respiratory system become affected later in the course of the disorder [6]. Within this consideration it should be borne in mind that perhaps as much as 50% loss of muscle fibres occurs prior to the development of symptoms. This is due to the normal reserve in muscle strength and the compensatory processes which take place in the early stages [7].

Exercise intolerance in the other dystrophinopathies

Following the discovery of dystrophin, a number of variant phenotypes from the classical DMD and BMD forms have been recognised in patients with lesser degrees of dystrophin deficiency. These include patients with weakness confined to the quadriceps muscles, patients with a clinical phenotype resembling limb-girdle dystrophy or spinal muscular atrophy, as well as patients with myoglobinuria or with myalgia, fatigue and muscle cramps or even some patients with asymptomatic elevation of the serum creatine kinase (CK). Such symptoms may also occur in female carriers, particularly those with dystrophin levels less than 50-60% of normal, and may be associated with some degree of muscle weakness and more frequently with elevation of the serum CK level (Table 1). Exercise intolerance in the manifesting carrier is due to outfall of muscle fibres and may be of a minor degree.

In patients with cardiomyopathy as a result of mutations in the dystrophin gene, exercise intolerance is mainly due to the impaired cardiac function which results from the reduced amounts of dystrophin reflected as diminished force of contraction. In heart failure, exercise intolerance is clearly related to reduced cardiac output and circulatory deficiency. In some patients, although the cardiomyopathy is the dominant clinical feature, varying degrees of muscle weakness within the limb girdle groups may be detected.

In the very mild forms of dystrophinopathy with myalgia and cramps it is presumed that the slight reduction in size of the dystrophin molecule and its relationship to the dystrophin-associated glycoproteins of the muscle membrane predisposes to entry of calcium which then accounts for the painful muscle contractions. Electrically silent but painful contractures as they occur in McArdle's disease, are not usually a feature of the dystrophinopathies.

Microscopic myoglobinuria is extremely common in all of the dystrophinopathies but is a special feature of DMD and the other disorders to a lesser extent. Myoglobinuria is usually a post exercise phenomenon. On occasions, myoglobinuria may be massive and lead to renal complications.

Conclusion

Exercise intolerance is the dominant feature of the symptomatology of the dystrophinopathies. The progressive nature of the disorder, especially in DMD and BMD, with continuous loss of muscle fibres and their replacement by fat and fibrous tissue, underlies the progression of symptoms. In DMD the boy becomes wheelchair bound at the ages of 11 or 12 years. In BMD ambulation may be retained for several decades. The clinical features and corresponding symptoms in BMD are similar to those for DMD but are less severe. In DMD, the disorder in every boy is almost the same. This stereotyping of the phenotype is due to total absence of dystrophin. However, in BMD where varying amounts of dystrophin may be present, the clinical features are variable and are correspondingly more or less severe depending upon the amount of dystrophin present.

In DMD, because of the rapid development of muscle weakness and consequent immobilisation, a host of secondary changes develop such as contractures of limbs, joints and scoliosis which greatly aggravate the clinical disorder. Scoliosis contributes to respiratory complications compounding the weakness of the intercostals, accessory muscles and diaphragm which occurs in the final stages of the disease. This involvement causes death due to intercurrent respiratory infections or simply respiratory failure and lack of oxygen.

Unlike DMD in BMD, because the disorder is less severe, the cardiomyopathy which accompanies all of the dystrophinopathies becomes more apparent in the later stages and death is due to congestive heart failure. Exercise intolerance in the later stages of BMD is due to a combination of heart failure and muscle weakness.

Appreciation of exercise intolerance in the dystrophinopathies assists in the planning of treatment and rehabilitation for the disorders.

Summary

The dystrophinopathies are a group of clinicopathological disorders resulting from mutations of the dystrophin gene on the X chromosome. The human dystrophinopathies are: Duchenne muscular dystrophy (DMD), Becker muscular dystrophy (BMD), manifesting female carriers, myocardopathies with or without minor limb girdle weakness, patients with myalgia and cramps and asymptomatic creatine kinase aemias. Dystrophin is a large cytoskeleton protein related to the spectrin family and has a molecular weight of 427kDa. Its main function is believed to be to strengthen the muscle membrane preventing disruption during contraction. Total absence of dystrophin produces the severe DMD phenotype, while varying quantities of truncated protein result in BMD and the other milder forms. Exercise intolerance in the human dystrophinopathies is the result of impaired muscle function due to the underlying muscle lesions. The chief symptom is muscular weakness which mainly involves the larger limb girdle muscles. In DMD, weakness begins at 5 years of age and progresses until the boy becomes wheelchair bound at the age of 11 or 12 years. In the early stages of the disorder, weakness is associated with reduced stamina and fatigue. Less common symptoms are stiffness or rigidity and cramps. Fortunately, pain is not a feature of DMD *per se* but may occur as the result of postural deformities. In the later stages, exercise intolerance is also aggravated by respiratory insufficiency resulting from scoliosis and diaphragmatic weakness which is the common cause of death. Exercise intolerance in BMD is similar to DMD but less severe. Death in BMD is due to cardiac failure. In DMD the cardiac lesions are usually silent due to the patients lack of mobility. In the dystrophin deficient myocardiopathies, exercise intolerance may be due to disordered cardiac rhythm and to congestive heart failure. In the milder and asymptomatic dystrophinopathies, sudden severe exercise may precipitate acute muscle necrosis with myoglobinuria which, if massive, may cause renal failure.

References

1. Kakulas BA (1996) The differential diagnosis of the human dystrophinopathies and related disorders. Current Opinion in Neurology 9:380-388
2. Vignos PJ (1979) Rehabilitation in the Myopathies. In: Vinken PJ, Bruyn GW (eds) Handbook of Clinical Neurology. Diseases of Muscle Part II. Elsevier North-Holland Publishing, chap 13, pp 457-500
3. Emery AEH (1993) Duchenne Muscular Dystrophy. 2nd edn, Oxford University Press
4. Mastaglia FL, Kakulas BA (1996) X linked dystrophies: clinical and pathological features, In: RJ Lane (ed). Handbook of Muscle Diseases. Marcel Dekker, New York, chap 17, pp 201-221
5. Kakulas BA (1996) The spectrum of dystrophinopathies. In: RJ Lane (ed) Handbook of Muscle Diseases. Marcel Dekker, New York, chap 19, pp 235-244
6. Kakulas BA, YuPu G (1986) Quantitative assessment of muscle lesions in 19 Duchenne necropsies. Muscle and Nerve 5S (Abst 39) 23:206
7. Kakulas BA, Adams RD (1985) Diseases of muscle: Pathological Foundations of Clinical Myology. 4th edn, Harper and Row Publish, Philadelphia

Dystrophinopathies and CPT Deficiency in 75 Patients with Exertional Heat Stroke and Rhabdomyolysis

D. Figarella-Branger, M. Aubert, G. Kozak-Ribbens, N. Bianco, C. Bartoli, M.F. Montfort, O. Deslangles, J. Pouget, P.J. Cozzone, G. Serratrice , J.F. Pellissier

Exertional rhabdomyolysis and exercise intolerance are usually due to muscle metabolic dysfunction (glycogenosis and carnitine palmitoyl transferase deficiency [1]). However, exercise intolerance may be secondary to dystrophin abnormalities [2-6].

In addition, some patients with exercise intolerance may suffered from Brody's syndrome, a rare syndrome where the sarcoplasmic reticulum calcium ATPase (Serca) acitivity in muscle is reduced due to a deficiency in the fast isotype (SERCA 1) [7]. In some patients, the immunoreactive SERCA 1 protein is absent. These patients have usually characteristic clinical and electromyographic features.

Exertional heat stroke (EHS) is a rare syndrome occuring after an intense effort in healthy young adults who have no apparent clinical weakness or biological abnormality. It is characterized by hyperthermia above 40°C, neurological signs and symptoms and rhabdomyolysis [8]. Patients suffering from exertional rhabdomyolysis or EHS may have abnormal *in vitro* contracture tests such as malignant hyperthermia susceptible patients [9]. Moreover, impairement of calcium adenosine triphosphatase has been reported in a patient presenting exertional rhabdomyolysis [10].

For ten years, in Marseille, muscle biopsies have been done in a large population of patients presenting EHS or exertional rhabdomyolysis to perform *in vitro* contracture tests.

We have searched for CPT deficiency, dystrophinopathy, sarcoglycanopathy and calcium adenosine triphosphatase deficiency in these patients.

Materials

Selection of patients

75 patients were included in this study. It encompassed, 60 patients, all military, who have presented EHS (59 males, 1 female; age < 30 years);

and 15 military patients suffering from exertional rhabdomyolysis (13 males and 2 females; age < 30 years). CK levels at rest were available in 37/60 patients with EHS. They were normal in 23, increased in 14 but never higher than 500 (Nl < 130). Two patients with exertional rhabdomyolysis had CK levels at rest higher than 1000.

Methods

Muscle biopsies

In all patients, muscle biopsies were carried out under local anaesthesia in the biceps brachialis. In each case, an interval of at least 3 months was left between the attack (EHS or exertional rhabdomyolysis) and the muscle biopsy. Fragments of muscle necessary for *in vitro* contracture tests, in accordance with the European protocol [11] and those required for morphological, immunohistochemical and biochemical analysis of the muscles were removed.

For the morphological and biochemical analysis, the samples were frozen in isopentane cooled by liquid nitrogen. Histochemical techniques were carried out on serial sections according to the previously reported methods [12].

Antibodies

Monoclonal antibodies directed against the different regions of dystrophin, sarcoglycans (Dag 35, 43, 50), Serca 1 and Serca 2, and utrophin (NCL DRP2) were purchased from Novocastra Laboratories. Anti-Dys 1, Dys 2 and Dys 3 antibodies react with the rod domain, the C-terminal and the N-terminal region of the protein, respectively. They were used at appropriate dilution for immunohistochemistry. For immunoblot analysis, anti-Dys 1 and Dys 2 were used at 1/180 dilution, respectively. Immunoblotting study with anti-Dys 3 antibody was not done.

Immunohistochemistry

Automated immunohistochemistry was performed on serial frozen section (4 µm thick) with avidin-biotin-peroxidase complex on Ventana 320 device [13], Ventana kits (Ventana systems, Strasbourg, France) including amino-ethyl carbazol reagent. Sections were counterstained with hematoxylin, dehydrated and mounted in glycergel.

Western blot analysis

Immunoblot analysis with anti-dystrophin antibodies was performed in 16 and 5 patients suffering from EHS and exertional rhabdomyolysis, respectively, and in one control muscle. In these cases, immunoblot was done because of lack of staining with at least one anti-dystrophin antibody or slight irregular pattern with all antibodies.

Muscle extracts were prepared according to the technique previously described by Dechesne et al. [14]: 10 µl of each sample was applied in duplicate onto a gradient resolving SDS - polyacrylamide gels. Separated proteins were electroblotted onto nitrocellulose sheets (Amersham, Chicago, Ill).

The sheets were then processed according to the previously described method [15]. They were probed with anti-Dys 1 and anti-Dys 2 antibodies.

DNA analysis

Multiplex PCR analysis was performed on DNA extracted from the two muscle biopsy samples presenting an abnormal dystrophin expression by Western blot analysis. We used 18 primers from exons 4, 8 + 9, 12, 17, 19, 44 and 512 [16], as well as exons 3, 6, 13, 43, 47, 50, 52 and 60 and the muscle promoter [17].

Isotope exchange assay for CPT measurement

A fraction of each frozen muscle specimen (20 mg) was used for CPT measurement. The isotope exchange method of Norum was used to quantitate CPT [18]. This method mesures the reversible incorporation of labeled carnitine into a pool of palmitoyl-carnitine (CPT 1 and CPT 2).

CPT deficiency was assessed when residual activity was less than 30% of normal value. Partial CPT deficiency was diagnosed when residual activity was between 30% and 50% of normal value.

Results

Histopathological findings

Among the 60 patients suffering from EHS, 46 patients had normal biopsies, 10 mild myopathic changes or rhabdomyolysis (including fibers of various size, centrally located nuclei, occasional necrotic and regenerative fibers), 2 multiminicores in type I muscle fibers, 1 type II atrophy and 1 increase of glycogen content.

Eight patients with exertional rhabdomyolysis had normal muscle biopsy, 5 mild myopathic changes or rhabdomyolysis, 1 type II atrophy and 1 a microvascularitis.

CPT measurements

CPT measurements in muscle were abnormal in 50% and 30% patients suffering from EHS and exertional rhabdomyolysis, respectively.

Three patients with EHS have CPT deficiency and 27 partial CPT deficiency.

One patient presenting exertional rhabdomyolysis had CPT deficiency, and 4 had partial CPT deficiency. Patients presenting abnormal CPT activity had normal muscle biopsy or not (Fig. 1)

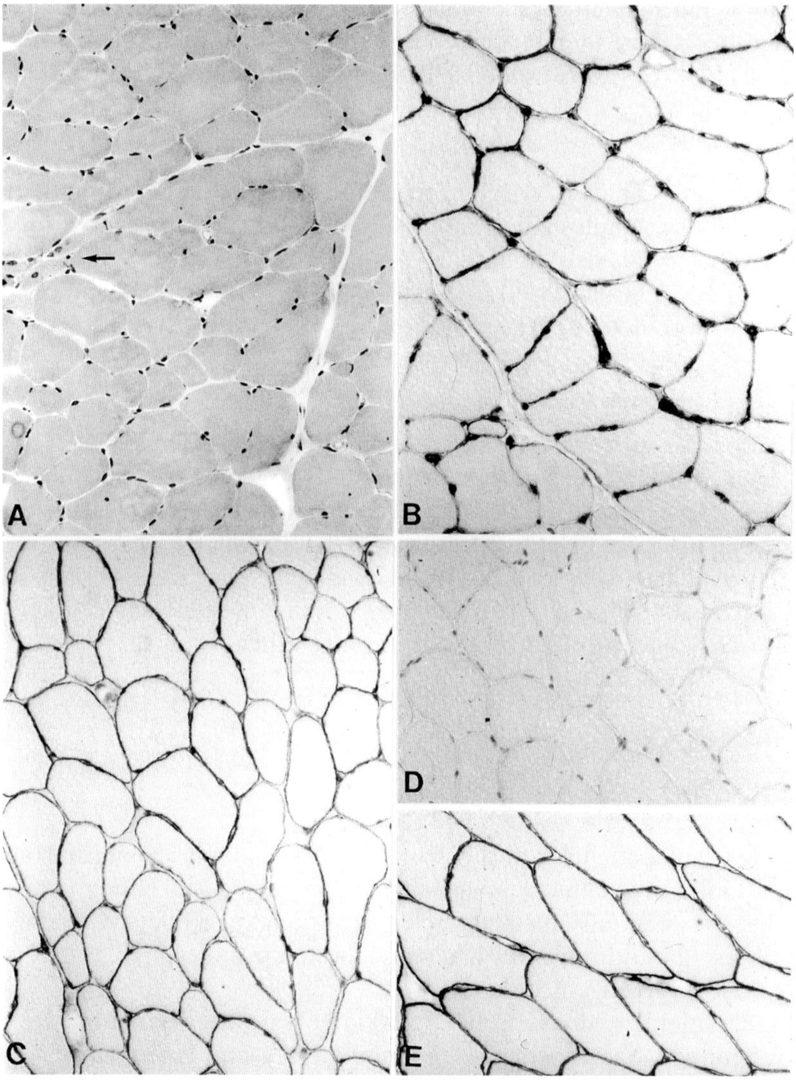

Fig. 1. Pathological changes and immunohistochemical detection of dystrophin (C, D) and utrophin (B) in one patient presenting an exertional rhabdomyolysis.
A. Fibers size variation and occasional regenerative fibers (*arrow*); B. Abnormal utrophin expression on cell surface membranes; C. Variation in labeling intensity with anti-Dys 2 antibody; D. Lack of anti-Dys 3 immunoreactivity (as compared with one control, E). A x 180; B-E x 200

Dystrophinopathy, sarcoglycanopathy and calcium adenosine triphosphatase deficiency

Sarcoglycanopathy or calcium adenosine triphosphatase deficiency were never observed.

Dystrophinopathy did not occur in patients suffering from EHS. In contrast, 2 out of 15 patients presenting exertional rhabdomyolysis had a dystrophinopathy. Both had high CK level at rest (> 1000) and mild myopathic changes at muscle biopsy. Immunohistochemistry showed irregular patterns of immunoreactivity with all anti-dystrophin antibodies in one case, lack of anti-Dys 1 and Dys 3 immunostaining in the other case (Fig. 1). Immunoblotting studies in the remaining cases were normal.

Anti-utrophin immunoreactivity was abnormal in both cases; it decorated the membrane of almost all muscle fibers. Immunoblot analysis revealed in both cases low amount of dystrophin of normal molecular weight (Fig. 2). PCR multiplex analysis of the dystrophin gene did not reveal any deletion.

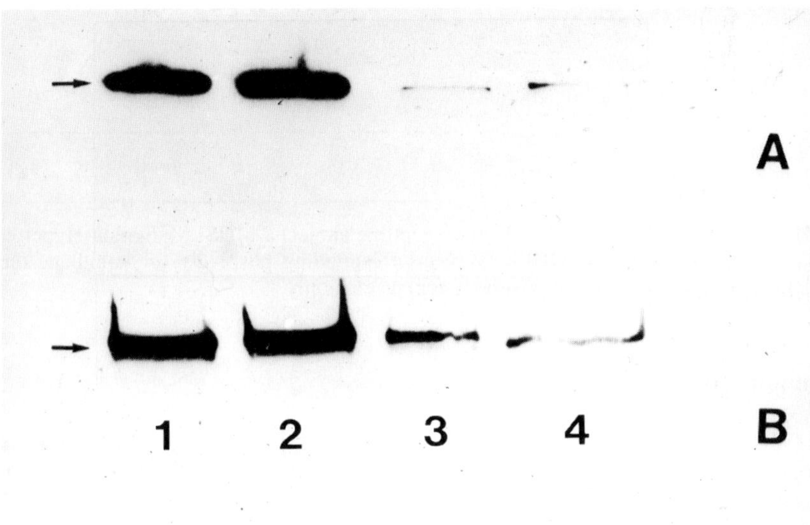

Fig. 2. Immunoblot analysis of dystrophin from normal muscles (lanes 1 and 2) and patients (lanes 3 and 4), respectively. Immunoblot analysis of Dys 1 (A) and Dys 2 (B) showing weak Dys 1 and Dys 2 expression in patients. Dystrophin molecular weight is normal

Correlations between histopathological findings, *in vitro* contracture test and CPT measurements

Among patients with EHS, 38 had normal *in vitro* contracture tests (MHN), 16 had abnormal *in vitro* contracture tests for both halothane and caffeine and were assumed to be susceptible for malignant hyperthermia (MHS), and 6 had abnormal contracture tests for halothane only (MHEh).

Among patients suffering from exertional rhabdomyolysis, 9 had normal *in vitro* contracture tests (MHN), 5 were MHS and one MHEh.

As previously reported [15] no correlation was observed between histopathological findings and *in vitro* contracture test results (Table 1) nor CPT measurements in muscle (Table 2). In addition CPT deficiency was not related to *in vitro* contracture test results (Table 3).

Table 1. Correlations between histopathological findings and *in vitro* contracture tests results

Histological features	Exertional heat stroke n = 60			Exertional rhabodomyolysis n = 15		
	MHN n = 38	MHEh n = 6	MHS n = 16	MHN n = 9	MHEh n = 1	MHS n = 5
Normal biopsy n = 54	30	3	13	5	1	2
Rhabodmyolysis and mild myopathic changes n = 15	5	2	3	3	0	2
Multiminicores n = 2	1	1	0	0	0	0
Miscellaneous n = 4	2	0	0	1	0	1

MHN: Malignant Hyperthermia Non susceptible subjects; MHS: Malignant Hyperthermia Susceptible subjects; MHEh: Subjects Responding positively for halothane only; MHEc: Subjects Responding positively for caffeine only

Comments

In this study, a dystrophinopathy was observed in two out of 15 patients suffering from exertional rhabdomyolysis but in no patients presenting EHS. It is likely, since 60 patients were included in this study, that EHS does not reveal dystrophinopathy. In contrast, as previously suggested [2-4, 6], exertional rhabdomyolysis may reveal dystrophinopathy. Interestingly, as pointed out previously, patients presenting a dystrophinopathy have high CK levels at rest and mild myopathic changes at muscle biopsy. We believe that search for dystrophinopathy in these patients is mandatory. Immunohistochemistry may be informative if it shows lack of anti-dystrophin immunoreactivity with anti-Dys 1 and/or anti-Dys 3 antibodies or abnormal immunostaining with all anti-dystrophin antibodies [6]. Moreover, if abnormal anti-utrophin immunoreactivity on the membrane of muscle fibers is observed, diagnosis of dystrophinopathy is almost certain. However, immunoblot analysis is required to confirm the diagnosis. In the present study, it demonstrates that the dystrophin molecule was normal in size but the total amount of the protein was strikingly decreased. We failed to observe dystrophin gene deletion by PCR multiplex analysis. The same results were reported previously and it is likely that dystrophin gene deletion in patients presenting same clinical features [6] does not occur when a decreased expression of a normal molecular sized dystrophin is observed.

Besides, this study shows that CPT deficiency frequently occur in patients suffering from EHS or exertional rhabdomyolysis: 50% and 30% of patients with EHS and exertional rhabdomyolysis, respectively, have abnormal CPT content in their muscles. The CPT enzyme consists of two immunologically distinct membrane bound enzymes CPT-I and CPT-II [19]. CPT-I is associated with the outer mitochondrial membrane, and is specifically inhibited by malonyl-COA [20] whereas CPT-II is located in the inner mitochondrial membrane and is insensitive to malonyl-COA [21].

Table 2. Correlations between histopathological findings and CPT measurements

Histopathological features	Exertional heat stroke n = 60			Exertional rhabdomyolysis n = 15		
	CPT deficiency ≤ 30%	CPT deficiency ≤ 50%	Normal CPT > 50%	CPT deficiency ≤ 20%	CPT deficiency ≤ 50%	Normal CPT > 50%
Normal biopsy n = 54	1	22	23	1	3	4
Rhabdomyolysis, mild myopathic changes n = 5	2	4	4	0	1	4
Miscellaneous n = 4	0	1	1	0	0	2

CPT: Carnitine Palmitoyl Transferase

Although the method we used to measure CPT content did not allow us to discriminate between CPT I and CPT-II, it is likely that the patients studied here suffered from CPT-II deficiency. In fact, CPT-II deficiency classically occurred in young adults, caused recurrent episodes of fasting or exercise induced myoglobinuria [1, 22]. CPT deficiency has been previously reported in one patient susceptible for malignant hyperthermia [23]. In our series, partial CPT deficiency was frequently observed. Genetic studies however are required to search for CPT deficient mutation in these patients [24, 26].

Besides, some patients with low CPT measurements in their muscles had also abnormal *in vitro* contracture tests. It has been reported that palmitoyl carnitine, the substrate which is elevated in CPT II deficiency, specially activates the calcium release channel in the sarcoplasmic reticulum of muscle [27]. Wether CPT deficiency is primitive or secondary to a deregulation of sarcoplasmic calcium cannot be assessed. Genetic studies are in progress to try to answer this question.

At last, in the population studied, sarcoglycanopathy or calcium adenosine triphosphatase deficiency were not observed and it is unlikely that such defects may induce EHS. Because of the low number of patients studied presenting exertional rhabdomyolysis, further studies are required to rule out such alterations in the triggering off of exertional rhabdomyolysis.

Table 3. CPT deficiency and *in vitro* contracture tests

CPT measurement	MHN	MHEh	MHS
CPT < 30%			
EHS n = 3	1	0	2
ER n = 1	0	0	1
CPT < 50%			
EHS n = 27	17	4	6
ER n = 4	3	0	1

CPT: Carnitine Palmitoyl Transferase; EHS: Exertional Heat Stroke; ER: Exertional Rhabdomyolysis

Acknowledgements

This work was supported by the Association Francaise contre les Myopathies and the Direction de la Recherche, des Etudes et de la Technologie.

References

1. Zierz S (1994) Carnitine palmitoyltransferase deficiency. In: Enegl AG, Franzini-Armstrong C (eds) Myology. 2nd edn, Mc Graw-Hill, New York, pp 1130-1187
2. Gospe SM, Lazaro RP, Lava NS, Grootscholten BS, Scott MO, Fischbeck KH (1989) Familial X-linked myalgia and cramps: a non progressive myopathy associated with a deletion in the dystrophin gene. Neurology 39:1277-1280
3. Gold R, Kress W, Meurers B, Meng G, Reichmann H, Müller CR (1992) Becker muscular dystrophy: detection of unusual disease courses by combined approach to dystrophin analysis. Muscle Nerve 15:214-218
4. Doriguzzi C, Palmucci L, Mongini T, Chiado-Piat L, Restgno G, Ferrone M (1993) Exercise intolerance and recurrent myoglobinuria as the only expression of Xp21 Becker type muscular dystrophy. J Neurol 240:269-271
5. Malapert D, Recan D, Leturq F, Degos JD, Gherardi RK (1995) Sporadic lower limb hypertrophy and exercice induced myalgia in a woman with dystrophin gene deletion. J Neurol Neurosug Psychiatry 59:552-554
6. Figarella-Branger D, Baeta Machada AM, Putzu GA, Malzac P, Voelckel MA, Pellissier JF (1997) Exertional rhabdomyolysis and exercise intolerance revealing dystrophinopathies. Acta Neuropathol 94:48-53
7. Karpati G, Charuk J, Carpenter S, Jablecki C, Holland P (1986) Myopathy caused by a deficiency of Ca^{2+} adenosine triphosphate in sarcoplasmic reticulum (Brody's disease). Ann Neurol 20:38-49

8. Aubert M, Petrognani R, Deslangles O (1993) Exertional heat stroke. In: Aubert M, Borsarelli J, Kozak-Ribbens G, Khambatta HJ (eds) Malignant hyperthermias. Normed Verlag, pp 5-13
9. Kozak-Ribbens G (1993) Investigations and diagnosis of malignant and exercise hyperthermias. In: Aubert M, Borsarelli J, Kozak-Ribbens G, Khambatta HJ (eds) Malignant hyperthermias. Normed Verlag, pp 44-62
10. Poels PJE, Wevers RA, Braakhekke JP, Benders AAGM, Veerkamp JH, Joosten EMG (1993) Exertional rhabdomyolysis in a patient with calcium adenosine triphosphatase deficiency. J Neuro Neurosurg Psychiatry 56:823-826
11. The European malignant hyperpyrexia group (1984) A protocol for the investigation of malignant hyperpyrexia. Br J Anaesth 56:1267-1269
12. Dubowitz V, Brooke MH (1973) Muscle biopsy: a modern approach. Saunders, London, pp 20-23
13. Grogan TM, Casey TT, Miller PC (1993) Automation of immunohistochemistry. Adv. Pathol Lab Med 6:253-283
14. Dechesne CA, Leger JOC, Bouvagnet P, Mairhofer H, Leger JJ (1985) Local diversity of myosin expression in mammalian atrial muscle. Circ Res 57:767-775
15. Figarella-Branger D, Kozak-Ribbens G, Rodet L, Aubert M, Borsarelli J, Cozzone PJ, Pellissier JF (1993) Pathological findings in 165 patients explored for malignant hyperthermia susceptibility. Neuromusc Disord 3:553-556
16. Chamberlain JS, Gibbs RA, Ranier JE, Nguyen PN, Caskey CT (1988) Deletion screening of the Duchenne muscular dystrophy locus via multiplex DNA amplification. Nucleic Acid Res 16:11141-11156
17. Beggs AH, Koening M, Boyce FM, Kunkel LM (1990) Detection of 9-8% of DMD/BMD gene deletions by polymerase chain reaction. Hum Genet 86:45-48
18. Norum K (1964) Palmityl-CoA carnitine palmitoyl-transferase. Biochem Biophys Acta 89:95-108
19. Woeltje KF, Kuwajima M, Foster DW, McGarry JD (1987) Characterisation of the mitochondrial carnitine palmitoyltransferase of the mitochondrial carnitine palmitoyltransferase enzyme system. II. Use of detergents and antibodies. J Biol Chem 262:9822-9827
20. Declerq PE, Falck JR, Kuwajima M, Tyrninski H, Foster DW, McGarry JD (1987) Characterisation of the mitochondrial carnitine palmitoyltransferase enzyme system. I. Use of inhibitors. J Biol Chem 262:9812-9821
21. Woeltje KF, Esser V, Weiss BC et al (1990) Inter-tissue and interspecies characteristics of the mitochondrial carnitine palmitoyltransferase enzyme system. J Biol Chem 265:10714-10719
22. Di Mauro S, Di Mauro PMM (1973) Muscle carnitine palmitoyl-transferase deficiency and myoglobinuria. Science 182:929-931
23. Vladutiu GD, Hogan K, Saponara I, Tassini L, Conroy J (1993) Carnitine palmitoyl transferase deficiency in malignant hyperthermia. Muscle and Nerve 485-491
24. Taroni F, Verderio E, Dworzak F, Willems PJ, Cavadini P, DiDonato S (1993) Identification of a common mutation in the carnitine palmitoyltransferase II gene in familial recurrent myoglobinuria patients. Nature Genet 4:314-320
25. Verderio E, Cavadini P, Montermini L et al (1995) Carnitine palmitoyltransferase II deficiency: structure of the gene and characterization of two novel disease-causing mutations. Hum Mol Genet 4:19-29
26. Schaefer J, Jackson S, Taroni F, Swift P, Turnbull DM (1997) Characterisation of carnitine palmitoyltransferases in patients with a carnitine palmitoyltransferase deficiency: implications for diagnosis and therapy. J Neurol Neurosurg Psychiatry 62:169-176
27. El Hayek R, Valdivia C, Valdivia HH, Hogan K, Coronado R (1992) Palmitoyl carnitine: an endogenous modulator of the Ca^{++} release channel of skeletal muscle. Biophys J 61(abstr):137a

Muscular Metabolism and Contraction in Exertional Heat Stroke

G. Kozak-Ribbens, M. Aubert, D. Figarella-Branger, D. Bendahan, P.J. Cozzone

A number of similarities have been observed between malignant hyperthermia (MH) and exertional heat stroke (EHS) syndromes. Malignant hyperthermia is a pharmacogenetic syndrome appearing in predisposed subjects characterized by rigidity and accelerated metabolism of skeletal muscles leading to a dramatic increase in core temperature. MH occurs in susceptible subjects after exposure to triggering agents such as inhalational fluorinated anesthetics and depolarizing muscle relaxants [1-3]. Exertional heat stroke (EHS), also called exercise hyperthermia, is characterized by hyperthermia above 40°C, neurological signs and rhabdomyolysis [1,4]. This syndrome occurs during intense, extended physical effort not only in hot and humid climatic conditions. Patients are healthy young adults who have no apparent clinical or biological deficit. Since the hypothesis of a preexisting muscular disorder was raised in EHS patients, our objective was to investigate the etiology of EHS and the possible link between EHS and malignant anaesthetic hyperthermia.

We have systematically investigated patients with EHS in the same way as patients with MH syndrome [5, 6]. Since 1990, EHS patients have been explored in Marseille by a specialized multidisciplinar team which is now the national referral center for the investigations of military patients having suffered from EHS.

The procedure for the diagnosis of MH susceptibility is based on *in vitro* contracture tests on muscle biopsies according to the EMHG protocol [7]. Additional explorations were also conducted to probe morphological abnormalities which could be specific of a myopathy. When abnormalities were observed, further biochemical studies were performed. 31P Magnetic Resonance Spectroscopy (31P MRS) has been used to explore muscular energetic metabolism non invasively, for possible metabolic dysfunctions in search of metabolic dysfunction [8].

Patients

Investigations were conducted on 192 patients. Most of them were military personnel. All patients had presented a clear episode of exertional heat stroke. They were generally young (average age: 25 years, extremes: 19-55 years). Generally they never had been anaesthetized or had received no more than one general anaesthesia. Some patients suffered from recurrent EHS: 6 patients from 2 episodes and one from 3. The episode appeared frequently at the end of a 8 km commando march or a 10 km race. Contrary to the accepted idea that a hot environment is necessary, patients were frequently affected in a moderate or cool weather. The first symptoms were neurological disorders, mainly behavioral anomalies which could lead to unconsciousness (50% of patients). The core temperature which was increase by the sustained muscular work, could increased even more if the thermic charge was not correctly dissipated. The mean core temperature was 40.4°C. The value of blood CPK is a good index of rhabdomyolysis, although exercise itself induces a moderate increase of CPK. However, the mean value observed, 3415 (UI/l), was too high for being due to normal exercise. For a small number of patients with low CPK values, rhabdomyolysis was delayed. The most frequent complications were hepatic cytolysis and disseminated intravascular coagulation (DIC). Myoglobinuria was frequent. Only in 7 patients acute renal insufficiency was observed.

Methods

Muscle biopsy

Biopsies were excised from biceps brachii muscle for 80% of patients after at least a 3 month delay following rhabdomyolysis and to avoid regenerative processes that could modify biopsy testing. A similar delay was observed before exploration by 31P MRS [6].

Contracture tests

These tests investigate excitation-contraction coupling which is abnormal in MH muscles. Contracture tests proceed from measure of *in vitro* contractures produced by muscle bundles in the presence of increasing concentrations of caffeine and halothane. Halothane/caffeine contracture tests were performed according to the European Malignant Hyperpyrexia Group protocol [7].

Tests are positive if a 0.2g-contracture (threshold) appears for 0.5 to 2mM of caffeine and 0.5 to 2% v/v (0.44mM) of halothane. If only one test is positive, the patient is classified as equivocal (MHE). At present, only MHEh subjects are recognized as MH susceptible until a genetic discrimination allows to ascribe them a definitive status.

An additional contracture test was made in the presence of ryanodine. Ryanodine binds specifically to the sarcoplasmic reticulum calcium channel which is known to be defective in MH muscles [9]. Testing was positive when 1μM ryanodine induced a contracture during the 5 first minutes of application (onset time).

Histoenzymological examinations and carnitine palmitoyl transferase activity (CPT)

These investigations have been systematically performed together with contracture tests [10] (see also Figarella-Branger et al. in this issue).

31P MRS

This technique was applied to study the muscular energetic metabolism concurrently to halothane/caffeine contracture tests [6]. 31P MRS was excluded for a few patients with contraindications to MR exploration.

NMR spectra were recorded at 4.7 Tesla on a Bruker 4730 Biospec system equipped with a superconducting horizontal magnet (bore diameter 30 cm). A 50 mm diameter surface coil was positioned over the flexor digitorum superficialis muscle. Spectra were time-averaged over 1 minute and sequentially recorded during 3 minutes of rest, 3 minutes of exercise and 20 minutes of recovery. The protocol was applied a second time under ischemic conditions limited to exercise, by inflating a cuff over the maximal arterial pressure. Relative concentrations of phosphorylated metabolites were determined by integration of their respective resonances. Concentrations of the main phosphorylated metabolites as creatine phosphate (CP), inorganic phosphate (Pi), adenosine triphosphate (ATP) could be quantified and values of the main ratios as CP/Pi, CP/ATP, rates of breakdown and built up of CP were calculated. Intracellular pH was calculated from chemical shift of inorganic phosphate relative to phosphocreatine [11]. Taking into account 19 parameters, a MRS score of predictive MH susceptibility was calculated [12].

Results

Contracture tests

Among the 192 patients tested, 59 could be considered as MH susceptible (Fig. 1, Table 1). The criteria of EMHG protocol separate this group into 41 MHS for which halothane and caffeine contracture tests were both positive, and 18 MHEh for which only halothane test was positive. These patients received an identification card indicating their MH susceptibility, and their families were informed of a possible transmission of the risk.

Often, family investigations could be initiated to determine the type of genetic transmission (dominant or recessive). The status of MHEc (positive caffeine test) patients remains to be clarified until a genetic discrimination ascribe a definitive status to them.

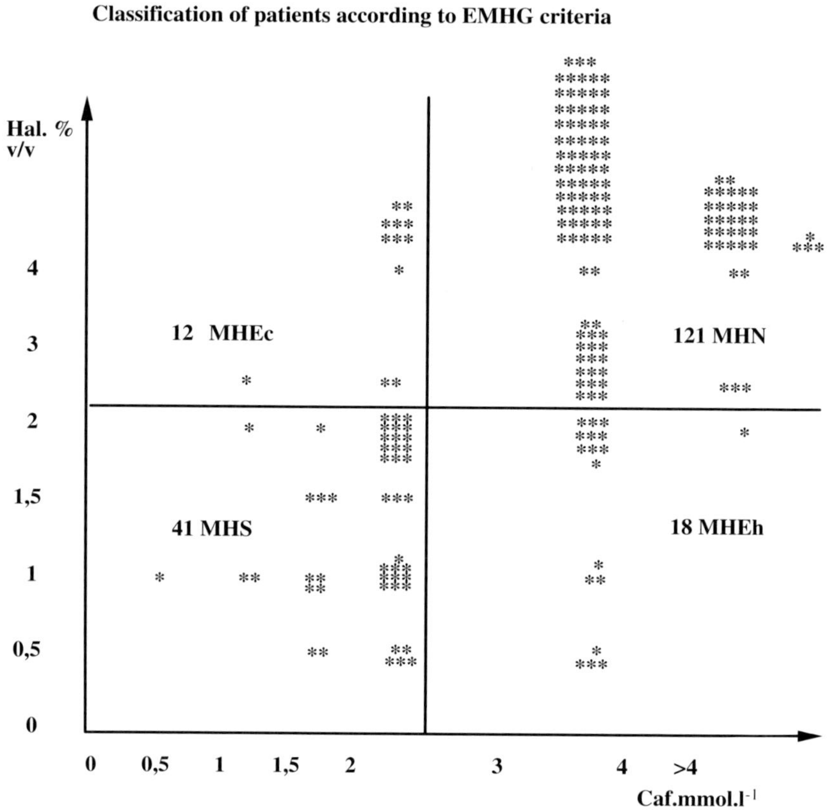

Fig. 1. Classification of the 192 patients investigated in Marseille. *Horizontal scale*: cumulatives caffeine concentrations; *vertical scale*: cumulatives halothane concentrations. Position of each patient is determined by at least a 0,2g contracture to a concentration of caffeine from 0.5 to 4mM, and of halothane fom 0.5 to 4%. To establish a diagnosis of MH a 0,2g contracture at a concentration of 2mM caffeine or less and 2% or less halothane is required: MHS patients are positioned in the left lower quarter; MHN or non susceptible patients in the upper right quarter; in the other two quarters, MHE or equivocal patients.

The 41 MHS patients could be separated into 2 groups with respect to results of ryanodine tests: those for which the onset time was in the range of MH susceptible patients belonging to MH families (66%), and those for which the onset time was over this range (35%). In the first case, a true link between MH and exertional heat stroke susceptibilities could be suspected.

Table 1. Contracture tests results for the 192 explored patients

Tests results	MHN	MHEh	MHEc	MHS
No. of patients	121	18	12	41
%	63%	9%	6%	21%

Relations between EHS clinical signs and contracture tests results

As it appears in Table 2, no statistical differences could be observed among the patients classified according to their contracture tests results and the mean increase of hyperthermia and CPK as well as in the frequency of hepatic cytolysis, renal insufficency or disseminated intravascular coagulation.

Table 2. Relationship between clinical signs and contracture tests results

Clinical signs	Hyper-thermia	CPK IU/l	Hepatic cytolysis	Renal insuff	DIC
MHN	40.4° C	4741	36%	7%	18%
MHEh	40.6° C	3464	25%	12%	12%
MHEc	40.7° C	2930	50%	16%	16%
MHS	40.2° C	4071	3%	6%	29%

Relations between histological and contracture tests results

Only a small number of patients exhibited morphological and enzymological abnormalities, as shown in Table 3. Anatomopathological examinations revealed some defined myopathies as: multiminicores disease [4], dystrophinopathy [4], mitochondrial myopathies [12], rhabdomyolysis "a minima" [13], capillaritis and microvascularitis [12], glycogen overload [14] and lipid overload [4]. All were clinically silent. No correlations could be drawn between EHS clinical symptoms and histological results as shown in Table 4.

Table 3. Relationship between histological and contracture tests results

Clinical signs	Normal biopsy	Discrete anomalies	Extended anomalies
MHN	70%	13%	17%
MHEh	50%	22%	28%
MHEc	58%	33%	9%
MHS	77%	20%	3%

31P MRS

Results of 31PMRS examination of muscular energetic metabolism showed several metabolic profiles:

- Normal metabolic profile. This profile was observed for approximately 10% of the patients. In these cases, the biopsy was not performed. Normal metabolic profiles have never been observed for patients with abnormal contracture tests.
- Abnormal profile identical to that of patients classified as MHS by contracture tests in MH families. For EHS patients, MHS profiles were observed in good agreement with the results of contracture tests. During exercise a significative metabolic dysfunction was observed, marked by a severe acidosis and an abnormal consumption of phosphorylated metabolites (CP and ATP). A delayed recovery to rest values of both acidosis and CP/Pi ratio was frequently observed [2, 6]. In the MHS patient group, MRS scores were in the MH susceptible patient range for 83% of the MHS patients.
- EHS profile. Abnormalities were clearly observed only during ischemic test, indicating that an impairment of oxygen supply was necessary to produce metabolic dysfunctions.
- Abnormal profiles with different patterns according to the defective metabolic pathway, corresponding to i) an abnormal increase or decrease in anaerobic glycolysis; ii) a reduced activity of oxydative metabolism; iii) anomalies of intracellular pH homeostasis, and iv) signs of a permanent rhabdomyolysis. For patients exhibiting these types of profiles, some well-defined myopathies were revealed also by anatomopathological investigations.

Table 4. Relationship between clinical signs and histological results

Clinical signs	Hyper-thermia	CPK IU/l	Hepatitis cytolysis	Renal insuff	DIC
Extended anomalies	40.5° C	6168	53 %	17 %	35 %
Discrete anomalies	40.7° C	10136	43 %	21 %	14 %

Discussion and comments

First, from these results on a large group of patients, a muscular origin of EHS could be confirmed in accordance with our previous reports [11, 15]. The same conclusion was drawn by Hackl et al. [14] and Hopkins et al. [13].

The second point is the link between MH and EHS since MH susceptibility was recognized by halothane/caffeine contracture tests in a significant proportion (30%) of the EHS patients. In these cases, molecular and genetic investigations were necessary to confirm MHS and EHS identity. MHS patients received a personal card and their families were informed of a possible transmission of the risk. MHS status for a patient implies to use special anaesthetic procedures without triggering agents if general anaesthesia is required. Also family investigations are generally undertaken. As for MH families, investigation of EHS families allows to reveal the MH susceptibility among family members, and consequently to prevent possible accidents related to anaesthesia.

Family investigations allow to determine the type of genetic transmission. MH susceptibility is linked to more than one gene [2, 16, 17], the first being the ryanodine receptor (RYR1) gene located on chromosome 19. The genetic testing would be initiated to probe the RYR1 mutations. A large number of patients were affected by silent muscular anomalities, whatever their MH status, suggesting that systematic histochemical investigations should be performed.

Taking into account the abnormal spectroscopic profiles in EHS patients, 31P MRS has proved useful as a first intention discriminative test for MH susceptibility and has brought valuable information in the search for associated myopathies.

Conclusion

Three main conclusions can be drawn:
1) in most cases, EHS is related to a "silent" muscle dysfunction
2) all patients who suffered from EHS episodes must be extensively explored to determine MH susceptibility and/or to reveal any possible myopathy;
3) for patients classified as MHS by contracture tests, family investigations should be carried out to search for a possible family susceptibility. The family investigation allows to prevent accidents during anaesthesia and to determine the type of genetic transmission.

Acknowledgements

All physicians from the military and university hospitals are acknowledged for patients referrals

This work was supported by grants from DRSSA (Direction de la Recherche du Service de Santé des Armées), APHM (Assistance Publique des Hopitaux de Marseille), CNRS (Centre National de la Recherche Scientifique), AFM (Association Française contre les Myopathies) and PHRC (Programme Hospitalier de Recherche Clinique).

References

1. Kozak Reiss G, Coursanges M, Aubert M (1991) Hyperthermies malignes. Editions Techniques. Encycl Med Chir (Paris-France), Techniques Chirurgicales – Anesthésie Réanimation, 364 12 E 10, 18p
2. Mac Lennan DH, Phillips MS (1992) Malignant hyperthermia. Science 256:789-794
3. Serratrice G, Kozak-Ribbens G (1996) Hyperthermie Maligne. Encycl Med Chirg, Neurologie, I. Médicaux 171:A-50

4. Aubert M, Petrognani R, Deslangles O (1993) Exertional Heat Stroke. In: Aubert M, Borsarelli J, Khambatta HJ, Kozak-Ribbens G (eds) Malignant Hyperthemias. Normed verlag Bad Homburg Englewood Madrid, pp 52-71
5. Kozak Reiss G, Borsarelli J (1990) "A rapid and extensive investigation of MH patients in Marseille by a pluridisciplinary integrated structure". 7th International Workshop on Malignant Hyperthermia, Munich 22-23 september. J Neurol Sci 98S:518
6. Kozak-Ribbens G (1993) Investigation and diagnosis of malignant and exercise hyperthermias. In: Aubert M, Borsarelli J, Khambatta HJ, Kozak Ribbens G (eds) Malignant Hyperthemias. Normed verlag Bad Homburg Englewood Madrid, pp 44-62
7. The European Malignant Hyperpyrexia Group, Ellis FR et al (1984) A protocol for the investigation of Malignant Hyperpyrexia (MH) susceptibility. Br J Anaesth 56:1267-1273
8. Bendahan D, Confort-Gouny S, Kozak-Ribbens G, Cozzone PJ (1993) Investigation of metabolic myopathies by P-31 MRS using a standardized rest-exercise-recovery protocol: a survey of 800 explorations. MAGMA 1:91-104
9. Mickelson JR, Louis CF (1996) Malignant hyperthermia: excitation-contraction coupling, Ca^{2+} release channel and cell Ca^{2+} regulation defects. Physiol Rev 537-592
10. Figarella-Branger D, Kozak-Ribbens G, Rodet L, Aubert M, Borsarelli J, Cozzone PJ, Pellissier JF (1993) Pathological findings in 165 patients explored for Malignant Hyperthermia susceptibility. Neuromusc Disord 3(5/6):553-556
11. Legros P, Jehenson P, Gascard JP, Kozak Reiss G (1992) Long term relationship between acute rhabdomyolysis and abnormal end energy phosphate metabolism potentiated by ischemic exercise. Medecine and Science in sports and Exercise 24:298-302
12. Bendahan D, Kozak-Ribbens G, Rodet L, Confort-Gouny S, Cozzone PJ (1998) 31 Phosphorus Magnetic Resonance Spectroscopy Characterization of muscular metabolic anomalies in patients with malignant hyperthermia. Anesthesiology 88:96-107
13. Hopkins PM, Ellis FR, Halsall PJ (1991) Evidence for related myopathies in exertional heat stroke and malignant hyperthermia. Lancet 338:1491-1492
14. Hackl W, Winckler M, Mauritz W, Sporn P, Steinbereithner K (1991) Muscle biopsy for diagnosis of malignant hyperthermia susceptibility in two patients with severe exercise-induced myolysis. Br J Anaesth 66:138-140
15. Kozak Reiss G, Gascard JP, Hervé P, Jehenson P, Syrota A (1988) Malignant and exercise hyperthermia: investigation of 73 subjects by contracture tests and P 31 NMR spectroscopy. Anesthesiology 69 (3A):415
16. Monnier N, Canet C, Krivosic Horber R, Adnet P, Reyford H, Kozak-Ribbens G, Rodet L, Cozzone PJ, Stieglitz P, Nivoche Y, Desmonts JM, Lunardi J (1994) Genetic linkage analysis of French Malignant Hyperthermia population. Minerva Anestesiol 60(S 3):25-30
17. Sudbrak R, Procaccio V, Klausnitzer M, Curran JL, Monsieurs K, Van Broeckhoven C, Ellis FR, Heytens L, Hartung EJ, Kozak Ribbens G, Heilinger D, Weissenbach J, Lehmann Horn, Mueller CR, Deufel T, Stewart AD, Lunardi J (1995) Mapping of a further malignant hyperthermia susceptibility (MHS) locus to chromosome 3Q13.1. Am J Hum Genet 56:684-691

Heroin, Methadone, HIV and Myalgia

J.-L. Gastaut

Myalgia can occur in active drug addicts as well as subjects undergoing withdrawal with methadone. HIV infection can also cause painful muscle complications.

Myalgia and drug abuse

Acute rhabdomyolysis

Acute rhabdomyolysis is usually observed in addicts who inject heroin intravenously. However, other substances have been implicated including cocaine, amphetamines, codeine, barbiturates, benzodiazepines and quinine. The most common cause of rhabdomyolysis in heroin addicts is compressive muscle injury inflicted during a sudden overdose coma [1, 2]. Depending on the duration and depth of coma, other factors including hypoxia, arterial hypotension and dehydration can exacerbate rhabdomyolysis and lead to renal failure [3]. Another possible cause of rhabdomyolysis is tetanus infection resulting from the use of non-sterile needles. In some cases however rhabdomyolisis is observed without overdose, coma or other apparent cause. Several hypotheses have been proposed for these unexplained cases including direct toxicity or allergy to heroin or other components in the injected substance.

Diagnosis of acute rhabdomyolysis after coma is usually made two or three days later with the appearance of delayed symptoms: myalgia and other muscle manifestations. Intensive local and diffuse muscle edema is often observed and may cause compression. In many cases ecchymosis and blisters are observed.

Muscle biopsy shows the presence of more or less extensive muscle fiber abnormalities ranging from minor changes to complete necrosis with hemorrhagic effusion.

Muscle involvement can be confirmed by laboratory tests showing high levels of muscle enzymes (mainly creatine kinase, but also SGOT, SGPT, LDH and aldolase). Myoglobulyn can also be detected in blood and urine but testing must be performed early since myoglobulinemia and myoglobinuria are transient.

A potential but preventable complication of rhabdomyolysis is acute renal failure with oligoanuria. Onset may be rapid. The duration of anuria can vary from a few hours to several weeks. Intense muscle catabolism occurs and can be demonstrated by high serum urea levels, acidosis, hyperphosphatemia, hyperkalemia and hyperuricemia.

The prognosis of acute renal failure in addicts with acute rhabdomyolysis is usually very good using current therapeutic modalities. Treatment is the same as for any other form of renal failure and daily hemodialysis may be necessary due to the intensity of catabolism.

Prevention of acute renal failure can be achieved by properly monitored fluid loading, rapid alkalization, adequate treatment of hypoxia, and administration of furosemide to increase diuresis.

The intensity of muscle lesions associated with acute rhabdomyolysis is often alarming. Fasciotomy may be required to relieve edema and resulting neurovascular complications. Other local complications are possible, including skin lesions and infection. Nevertheless, unlike other types of rhabdomyolysis, the overall prognosis is good with complete recovery of renal function and mild muscle damage. Appropriate physiotherapy can limit motor sequels.

Acute fibrotic myositis

Acute fibrotic myositis [1], can result from progressive obstruction of veins due to repeated subcutaneous or intramuscular drug injections. Edema and congestion of muscles promote and intensify episodes of painful acute myositis, subcutaneous or intramuscular abscess and cellulitis. Direct drug effects on muscle tissue have also been implicated. Electromyography shows alternating myogenic and silent zones. Evolution is stopped if the injections are stopped.

A few cases of fibrotic myopathy have also been described after long-term intramuscular injection of other substances including pentazocine [4, 5], meperidine [6], hydromorphine hydrochloride [7] and butorphanol [8]. Myopathy can lead to deformations due to fibrotic infiltration of the muscles underlying the site of drug injection.

Infectious disorders

Infectious disorders are also possible causes of myalgia in drug addicts. In such cases the limits of myositis and necrotizing fasciitis can be determined by ultrasound, gallium scintigraphy, CT-scan and MRI. Treatment relies mainly on parenteral administration of antibiotics after bacteriologic testing to identify the offending organism. Surgical drainage and debridment are sometimes necessary.

Myalgia and methadone

Methadone is a synthetic antagonist of opiate receptors that has been known for almost 50 years. It is used mainly as substitute product during withdrawal in drug addicts. Like heroin and opiates, methadone can lead to acute rhabdomyolysis similar to those that have been described before [9, 10]. Rare cases of fibrotic and calcified myopathy [11] have also been described with intramuscular injections of methadone. Such cases seem to be due to a generalized myotoxic effect of the drug leading to symmetrical and severe contracture in all four limbs.

Myalgia and HIV infection

About 25% HIV - infected patients complain of muscle pain. Regardless of the cause, myalgia can occur at rest, during exercise, or when pressure is applied to muscles. Pain can be disabling due to intensity or functional limitation. Diagnosis of muscle involvement depends on clinical findings, electromyography and blood levels of muscle enzymes. Identification of the cause depends on anamnesis and often on biopsy of the deltoid or quadriceps muscles. Management depends on conventional analgesic drugs and treatment of the cause.

Rhabdomyolysis

Rhabdomyolysis is rare but can occur at any stage of HIV infection [12]. It is associated with myalgia, motor deficits and elevated muscle enzyme levels (especially creatine kinase) with risk of renal failure. The clinical picture rarely involves major muscle weakness. Rhabdomyolysis can be caused by the virus itself either during primary infection or as an isolated or recurrent event during the course of infection [13]. It can also be due to toxic effects of drugs such as didanosine (Ddi) which is toxic for mitochondria, sulfamethoxazole-trimethoprime, sulfadiazine which probably triggers an allergic immune response, pentaminidine, or lamivudine [14]. Finally, late-stage rhabdomyolysis in full-blown AIDS may be due to muscle toxoplasmosis, staphylococcal septicemia, repeated infectious episodes, or multiple treatment.

Polymyositis or myopathy

Polymyositis, also known as HIV myopathy, may be associated with HIV infection and can occur at any time during the course of disease. The onset of manifestations involves a subacute proximal deficit associated with resting or exertional myalgia predominating in the thighs and, in some cases, with muscle edema. Polymyositis are most often observed in the middle or advanced stages of HIV infection [15, 16]. Clinical findings are identical to those observed in seronegative subjects with idiopathic polymyositis. Creatine kinase levels are slightly or mildly elevated (2 to 4 fold higher than normal). Electromyograms show myogenic findings associated with spontaneous activity: fibrillations and positive sharp waves. Muscle biopsy is very useful. Findings include necrotic muscle fibers, inflammatory lympho-monocyte infiltrates with a predominance of CD8 lymphocytes, microvascularitis, and presence in some cases of nemaline rod bodies, cytoplasmic bodies and mitochondrial abnormalities. Polymyositis is never due to direct infection of muscle fibers by the virus. They are due to a cytotoxic CD8 process affecting cells expressing the class 1 major histocompatibility complex [17]. Some patients respond to corticosteroid therapy.

AZT-induced myopathy

The risk of AZT or zidovudine-induced myopathy increases with high cumulative doses, usually at 200 g or more [15, 16, 18]. Myopathy usually occurs after about 6 months of treatment and is generally associated with amyotrophy of limb girdles. Myalgias are inconstant. The main diagnostic criteria is regression of symptoms in the weeks following discontinuation of AZT treatment or reduction in dosage. Muscle enzyme levels are only mildly elevated. In the cases in which muscle biopsy has been performed, results demonstrated characteristic signs of mitochondrial myopathy, notably with the presence of ragged red fibers. After symptoms have subsided, treatment with AZT can often be resumed without reappearence of muscle symptoms, but careful clinical and laboratory surveillance is necessary.

Pyomyositis

Pyomyositis is rare in HIV-infected patients. However, it must be considered in differential diagnosis of patients presenting fever associated with diffuse or local muscle weakness and sometimes muscle swelling. Diagnosis depends on the findings of muscle biopsy or of needle puncture guided by ultrasound, CT-scan or MRI. Blood cultures may be negative. The most frequent microbial agent involved is *Staphylococcus aureus* [19-21]. Treatment mainly involves administration of antibiotics but drainage may also be required in some cases.

Pyomyositis has also been attributed to other types of infectious agents including cytomegalovirus [22], *Mycobacterium tuberculosis* [23], *Mycobacterium avium intracellulare* [24], *Cryptococcus neoformans* [25] or *Toxoplasma gondii* [26]

Primary muscular lymphoma

Malignant non-Hodgkin lymphoma can develop primarily in skeketal muscle tissue [27] and presents as a rapidly growing intramuscular tumor associated with fever. Differential diagnosis with pyomyositis can be made by radiologic imaging and biopsy.

Recurrent myoglobinuria

Polymyositis itself rarely causes myoglobinuria. Isolated recurrent myogloninuria is possible but very rare [13]. Each episode is characterized by myalgia, weakness in the extremities, elevation of creatine kinase and brown urine. Muscle biopsy reveals necrosis with no inflammation or abnormality in glycolytic enzymes. The patient is asymptomatic between episodes. Myoglobinuria is uncommon and is not a poor prognostic factor for imminent onset of full-blown AIDS.

References

1. Pesce A, Cassuto JP, Quaranta JF (1988) Pathologie somatique du toxicomane. Masson Paris, p 132
2. Touzeau D, Jacquot C (1977) Les traitements de substitution pour les usagers de drogues. Arnette, Paris, pp 176-177
3. Pesce A, Cassuto JP, Quaranta JF (1988) Pathologie somatique du toxicomane. Masson Paris, p 165-167
4. Oh SJ, Rollins JL, Lewis I (1975) Pentazocine-induced myopathy. J Am Med Assoc 231:271-273
5. De Schepper AM, Degryse HR (1990) Imaging findings in a patient with pentazocine-induced myopathy. Am J Roentgenol 154:343-344
6. Aberfeld DC, Brenenstock H, Shapiro MS, Namba T, Grob D (1968) Diffuse myopathy related to meperidine addiction in a mother and daughter. Arch Neurol 19:384-388
7. Mastaglia FL, Argov Z (1981) Drug-induced neuromuscular disorders in man. In: Walton J (ed) Disorders of voluntary muscle. Churchill Livingstone, New-York, pp 873-875
8. Wagner JM, Cohen S (1991) Fibrous myopathy from butorphanol injections. Journal of Rheumatology 18:1934-1935
9. Genth E (1979) Medikamentinduzierte Muskelerkrankugen. Therapiewoche 29:6091-6099
10. Hojs R, Sinkovic A (1992) Rhabdomyolysis and acute renal failure following methadone abuse. Nephron 62:362
11. Carpintero P, Kindelan J, Lluch M, Mesa M (1996) Methadone-induced fibrous and calcified myopathy - a report of 2 cases. Acta Orthop Scand 67:404-406
12. Chariot P, Ruet E, Authier FJ, Levy Y, Gherardi R (1994) Acute rhabdomyolysis in patients infected by human immunodeficiency virus. Neurology 44:1692-1696
13. Younger DS, Hays AP, Uncini A, Lange DJ, Lovelace RE, Di Mauro S (1989) Recurrent myoglobinuria and HIV seropositivity: incidental or pathogenic association? Muscle Nerve 12:842-843
14. Mendila M, Walter GF, Stoll M, Schmidt RE (1997) Rhabdomyolyse unter antiretroviraler Therapie mit Lamivudin. Deutsche Medizinische Wochenschrift 122:1003-1006
15. Simpson DM, Bender AN (1988) Human immunodeficiency virus-associated myopathy: analysis of 11 patients. Ann Neurol 24:79-84
16. Grau JM, Masanes F, Pedrol E, Casademont J, Fernandez-Sola J, Urbano-Marquez A.: Human immunodeficiency virus type 1 infection and myopathy: clinical relevance of zidovudine therapy. Ann Neurol 34:206-211

17. Illa I, Nath A, Dalakas M (1991) Immunocytochemical and virological characteristics of HIV-associated inflammatory myopathies: similarities with seronegative polymyositis. Ann Neurol 29:474-481
18. Panegyres PK, Papadimitriou JM, Hollingsworth PN, Armstrong JA, Kakulas BA (1990) Vesicular changes in the myopathies of AIDS. Ultrastructural observations and their relationship to zidovudine treatment. J Neurol Neurosurg Psychiatry 53:649-655
19. Blumberg HM, Stephens DS (1990) Pyomyositis and human immunodeficiency virus infection. Southern Medical Journal 83:1092-1095
20. Thomas R, French MA (1991) Pyomyositis complicating the acquired immunodeficiency syndrome. A report of two cases with coexistent neutropenia. Medical Journal of Australia 154:481-483
21. Patel SR, Olenginski TP, Perruquet JL, Harrington TM (1997) Pyomyositis: clinical features and predisposing conditions. Rheumatology 24:1734-1738
22. Ho HW, Bailey R, Rhee JM, Vinters HV (1989) Neuromuscular pathology in patients with acquired immunodeficiency syndrome (AIDS): an autopsy study. J Neuropathol Exp Neurol 48:382
23. Lupatkin H, Brau N, Flomenberg P, Simberkoff MS (1992) Tuberculous abscesses in patients with AIDS. Clinical Infectious Diseases 14:1040-1044
24. Wolf RF, Sprenger HG, Mooyaart EL, Tamsma JT, Kengen RA, Weits J (1990) Non-tropical pyomyositis as a cause of subacute, multifocal myalgia in the acquired immunodeficiency syndrome. Arthritis & Rheumatism 33:1728-1732
25. Anders K, Steinsapir KD, Iverson DJ et al (1986) Neuropathologic findings in the acquired immunodeficiency syndrome (AIDS). Clin Neuropathol 5:1-20
26. Gherardi R, Baudrimont M, Lionnet F et al (1992) Skeletal muscle toxoplasmosis in patients with acquired immuno-deficiency syndrome: a clinical and pathological study. Ann Neurol 32:535-542
27. Chevalier X, Amoura Z, Viard JP, Souissi B, Sobel A, Gherardi R (1993) Muscle lymphoma in patients with AIDS: a diagnostic problem with pyomyositis. Arthritis Rheum 36:426-427

Cardiorespiratory Correlations in Exercise Intolerance

G. Nigro, L.I. Comi

Exercise intolerance can derive from muscular, haematological, cardiac and respiratory impairments.

When the muscular disorders cause a complete inability to perform any exercise, as in the chair-bound state of Duchenne or SMA, the role played by the cardiovascular or respiratory apparatus is very poor; viceversa, when the muscular disorder depresses only partially the capacity to produce work, an eventual reduced cardiorespiratory performance plays a very important role. In this event in fact, the increased muscular needs cause cardiac and respiratory troubles that share many clinical features, particularly shortness of breath and wheezing. The myocardial damage, frequently associated with muscular diseases, and secondary abnormalities of respiratory function can severely worsen the clinical picture. Furthermore, the response to the exercise do not depend exclusively from the cardiorespiratory performance but even from both neuronal and endothelial Nitric Oxide (NO), that contributes to the regulation of vasomotor tone [1].

In the case of cardiomuscular damage and reduced NO synthesis, the exercise produces increase of sympathoexcitatory effects and increase of pulmonary blood pressure, first step of the left heart failure often developing toward the pulmonary oedema.

The severity of the cardiorespiratory dysfunction in the exercise intolerance can be measured by means of respiratory and cardiac functional tests. Such an evaluation, performed in a very large population of cardiac and muscular patients, followed up to 10 years in the Cardiomyology and Genetic Section of the 2nd University of Naples, allowed us to subdivide in four different stages the clinical pictures of exercise intolerance in the muscular patients.

During exercise, total body oxygen consumption - linearly increasing with the intensity of the exercise - is equal to the product of cardiac output (heart rate x stroke volume) and the total arterial-venous oxygen content difference [2].

$$\text{Exercise} = \underset{\text{(watt)}}{\text{Muscular work}} = \underset{\text{(ml/min)}}{\text{Oxygen consumption}}$$

$$\underset{\text{(ml/min)}}{\text{Oxygen consumption}} = \frac{\text{Heart Rate x Stroke Volume}}{\text{Vascular resistance}} \text{ x } \underset{\text{(ml/100 ml)}}{(\text{a - vO}_2\text{ difference})}$$

Therefore, the response to exercise is complex and based on adaptations within several systems, such as muscles, nervous system and blood, with the aim of supplying more oxygen to the working muscles, and eliminating the product resulting from increased metabolic activity.

When the muscular work is impaired by severe myopathies causing complete inability to perform any exercise, as in the advanced stages of Duchenne dystrophy (DMD) or in severe spinal muscular atrophy (SMA), the role played by the cardiovascular apparatus in exercise intolerance is poor; on the contrary, when the muscular disorders only partially reduce the ability to produce work, an eventually depressed cardiorespiratory performance can be very important in determining the intolerance to exercise (see Table 1).

Table 1. Causes of exercise intolerance

Type of disorder	Example	Consequence
Nervous	Severe SMA	Inability to perform exercise
Muscular	Late stage of DMD	Inability to perform exercise
Haematological	Severe anaemia	Increased cardiorespiratory work / fatigue
Cardiac	Heart failure	Pulmonary oedema / dyspnoea / fatigue
Respiratory	Obstructive respiratory failure	Increased cardiac work / dyspnoea /fatigue
Metabolic	Reduced NO synthesis	Reduced capacity of cardiorespiratory apparatus in producing work / dyspnoea

In haematological disorders, particularly in severe anaemia, the arterial-venous oxygen difference/ml is lower; consequently, the heart rate and the stroke volume need to be raised, but when the anaemia is too severe, or any cardiac or respiratory trouble is present, the heart performance may not be able to compensate the reduced arterial-venous oxygen difference, and exercise intolerance occurs.

Exercise intolerance often derives from myocardial damage [3] - frequently associated with muscular diseases - when skeletal muscle capacity to produce work is only partially impaired. In these cases the myocardial damage is worsened by the taxing requirement of oxygen supply to the skeletal muscles moving with faulty movements. Such a situation can cause heart failure with increased pulmonary blood pressure and severe intolerance to exercise. Pulmonary oedema may occur in the advanced stages.

In advanced stages of muscular diseases, respiratory failure occurs following the hypoventilation caused by the actual restriction of movement of the lungs, when the breathing work is increased by kyphoscoliosis [4]. Furthermore, thoracic movement may be weakened as a result of the neuromuscular disorders. Respiratory failure can produce very severe consequences if associated with cardiac damage and/or with increased need of oxygen supply to the skeletal muscles.

Finally, the cardiorespiratory disorders and the accumulation of products deriving from metabolic activity in degenerative diseases are responsible for an impairment of the NO synthesis contributing to sustained sympathetic activation and increased vascular resistance. In such a situation cardiovascular performance diminishes and consequently exercise intolerance occurs.

The entity of cardiorespiratory dysfunction in exercise intolerance can be evaluated through many different respiratory and cardiac functional tests [5-7], among which the more frequently performed are the following (see Table 2):

Table 2. Functional tests

Respiratory function tests	• tests of respiratory mechanics • tests of gas exchange • tests of exercise tolerance
Cardiac function tests	• WHO exercise protocol • Bruce exercise protocol • steep exercise protocol • exercise thallium scintigraphy • pharmacological echocardiography

Respiratory function tests

Tests of respiratory mechanics

The degree of distension and the rate of airflow could be quantified directly by measurement of pulmonary compliance and airways resistance, but in practice the elastic behaviour of the system is more easily assessed by measurement of lung volumes (particularly vital capacity).

Tests of gas exchange

The efficiency of pulmonary gas-exchange is assessed by measurement of the carbon monoxide transfer factor, also known as carbon monoxide (CO) diffusing capacity. However, the CO transfer factor is a sensitive but non specific index of the integrity of the alveolar-capillary membrane.

Tests of exercise tolerance

At the simplest level an exercise test consists of walking the patient along a corridor (without or with a treadmill) or performing it by bicycle ergometer. In more sophisticated exercise tests, ventilation and oxygen consumption are also measured.

The patients in an advanced clinical stage of muscular diseases are able to perform an exercise test walking only without treadmill.

Cardiac function tests

Cardiac function tests are mostly exercise tolerance tests performed by different techniques or using different protocols (see Table 2). In order to be widely adopted an exercise protocol has to meet two basic criteria: 1) it must be standardised, and 2) it must be normalised.

For standardisation purposes every cardiologist must use and adhere to exactly the same exercise protocol, either on the bicycle ergometer or on the treadmill. For normalisation purposes, normal expected values, collected in a large group of normal subjects of different age and sex, should be available.

Personal data

Respiratory and cardiac functional tests were performed in a group of 867 patients affected by primary myopathies and/or cardiomyopathies (Table 3).

Table 3. Stages of the clinical course of myopathy

1st Initial	without evident failure of skeletal muscle function at rest
2nd Intermediate	with evident failure of skeletal muscle function but ability to perform Gowers manoeuvre
3rd Advanced	walking patients not able to perform Gowers manoeuvre
4th Chairbound stage	

Almost all patients had been followed for up to 10 years in the Cardiomyology and Genetics Section of the 2nd University of Naples. Among them we selected a group of 80 Becker patients, aged between 15 and 45 years, affected by both muscular dystrophy and cardiomyopathy. The muscular impairment was classified according to our proposal published in 1995 [8] (Table 4).

Table 4. Muscular and cardiac status of examined patients

Group	N° of Patients	Muscular impairment	Cardiac impairment
1	24	2nd stage	mild
2	19	3rdstage	mild
3	21	2nd stage	intermediate
4	16	3rd stage	intermediate

On the basis of the severity of cardiac and skeletal muscle impairment, the patients were subdivided into 4 groups (Table 5).

Patients with severe myopathies and/or severe cardiomyopathies, unable to perform any exercise, were not considered.

In the examined patients we considered the effects of exercise on ejection fraction (EF), dyspnoea on effort, artero-venous O_2 difference and on the rate of expiratory flux (Table 5).

Table 5. Results of exercise tolerance tests in Becker patients

Group	Ejection fraction below 50%	Dyspnoea	Remarkable increase of a-v O_2 difference	Reduced expiratory flux
1 (24)	2 (8.3%)	8 (33.3%)	9 (37.5%)	0
2 (19)	0	16 (84.2%)	19 (100%)	2 (10.5%)
3 (21)	18 (85.7%)	20 (95.2%)	20 (95.2%)	18 (85.7%)
4 (16)	15 (93,7%)	16 (100%)	16 (100%)	16 (100%)

The severity of the changes in the above reported parameters was very different in the four groups of patients examined.

The results summarised in table 5 show that ejection fraction depends only on cardiac status; the advanced myopathy, if isolated, can produce an increase but not a decrease in EF.

Dyspnoea on effort can be consequent both to myocardial damage and to the increased oxygen demand from skeletal muscles partially impaired.

The increase of arterio-venous O_2 difference on effort can be also due both to cardiac insufficiency and to the increased oxygen demand from partially impaired skeletal muscles.

The decrease in the rate of expiratory flux on effort is consequent to the onset of heart failure and to the successive initial pulmonary oedema. In this case an early obstruction of the airflow occurs.

In any case, the most evident expression of exercise intolerance in skeletal muscular patients is represented by dyspnoea on effort and increased artero-venous O_2 difference.

In the future, we are planning to investigate the role played by NO and TNF in exercise intolerance. In fact the decrease of NO and the increase of TNF seem to be present in situations of severe chronic muscular damage with consequent impairment of the cardiovascular function. The understanding of the role played by these two factors on the cardiorespiratory correlations in exercise tolerance in patients affected by skeletal myopathies, may explain many interesting aspects at the moment not yet completely clarified.

References

1. Owlya R, Vollenweider L, Trueb L et al (1997) Cardiovascular and Sympathetic Effects of Nitric Oxide Inhibition at Rest and During Static Exercise in Humans. Circulation 96(11):3897-3903
2. Sonnenblick EH, Ross J, Braunwald E (1968) Oxigen consumption of the heart. Am J Cardiol 22:328-36
3. Cowley AJ, Fullwood L, Stoinerk K, Hampton JR (1991) Exercise tolerance in patients with heart failure. Eur Heart J 12:50-54
4. Mc Porland C, Krishnan B, Wang Y, Callagher CG (1992) Inspiratory muscle weakness and dyspnoea in chronic heart failure. Am Reo Respir Dis 146:467-472
5. Dilsizian V, Borrow RO (1993) Current diagnostic techniques of assessing myocardial viability in patients with hibernating and stunning myocardium. Circulation 87:1-20
6. Gibson GJ (1995) Clinical Tests of Respiratory Function. 2nd edn, Chapman and Holl, London
7. Redwood DR, Rosing DR, Goldstein RE et al (1971) Importance of the design of an exercise protocol in the evaluation of patients with angina pectoris. Circulation 43:618-628
8. Nigro G, Comi LI, Politano L et al (1995) Evaluation of the Cardiomyopathy in Becker muscular dystrophy. Muscle & Nerve 18:283-291

Fibromyalgia: Symptom or Disease?

R.B. Layzer

The term "fibromyalgia" (FM) is currently used to describe patients with chronic, unexplained soft-tissue pain and tenderness, usually accompanied by numerous other unexplained symptoms such as fatigue, insomnia, headaches, bowel disturbances, anxiety and depression. Under other designations (fibrositis, muscular rheumatism, myofascial pain) the syndrome was extensively discussed in German and British literature from 1843 to 1950, and the diagnosis had a surge of popularity in the United States after 1970. Physical examination of FM patients shows no abnormality except for multiple sites of soft-tissue tenderness; previous descriptions of palpable nodules have not been confirmed. Laboratory tests show no consistent abnormality, including serum enzymes, EMG, muscle biopsy, muscle biochemistry, exercise physiology tests, and P^{31}-NMR spectroscopy. However, pain physiology tests have shown altered cerebral processing of pain information, with lower pain thresholds and reduced central inhibition. The lifetime history of major depression is 71%, 5 times the incidence in rheumatoid arthritis, yet only 25% of FM patients are currently depressed. These findings suggest that FM (like Chronic Fatigue Syndrome) is a cerebral disorder with a close link to depression, though it is not identical to depression and rarely responds dramatically to antidepressant drug therapy.

Fibromyalgia: symptom or disease?

Generalized muscle pain is one of the most common symptoms that patients bring to their physicians. Although a specific diagnosis can sometimes be made, in about one-half to two-thirds of cases no cause can be found, even after expert evaluation in a muscle disease clinic [1, 2]. In recent years, such patients have often been labelled with the diagnosis of fibrositis or primary fibromyalgia. How useful is that diagnosis? Does it refer to a specific disease, or does it merely designate a subjective complaint without objective components?

History of the concept of fibromyalgia

Simons [3] has provided a detailed historical account of the idiopathic muscle pain syndrome. The story begins in Germany in 1843, when Froriep described rheumatic patients with painful muscle hardenings (Musckelschwiele). Later German authors also reported the presence of localized areas of tenderness and hardening, although other physicians doubted that the hardenings were genuine. In 1919, Schade observed that the hardenings (Myogelose) persisted during general anesthesia. Biopsies of these nodules were reported to show a variety of pathological changes.

British authors seconded these observations. In 1900 Adler introduced the term "muscular rheumatism," which he attributed to interstitial myositis, and he thought that fat, fascia and nerves were also involved. Gowers [4], in 1904, coined the term "fibrositis", but he was referring to lumbago and sciatica, which he attributed to inflammation around the sciatic nerve. In the same year, Stockman [5] described muscle biopsy pathology in the syndrome of muscular rheumatism. Adopting the term "fibrositis," Llewellyn and Jones [6] described patients with generalized muscle pain, stiffness and fatigue, with or without tender nodules.

Initially, North American physicians expressed skepticism about the validity of the condition they knew as muscular or non-articular rheumatism. Hench [7] dismissed the nodules of fibrositis as "only accessible to the finger of faith," and Ellman and Shaw [8] attributed the syndrome to psychiatric causes: "The patient aches in his limbs because in fact he aches in his mind".

By the middle of this century, however, a new enthusiasm for fibrositis appeared in North American medical circles. In the early 1940's, Janet Travell (who later treated President Kennedy) popularized Kraus's concept of myofascial pain as a vicious circle, in which pain causes muscle spasm, which causes more pain. She treated so-called trigger-points by spraying the area with ethyl chloride and stretching the underlying muscle, a procedure also borrowed from Kraus [9]. In 1971 Smythe, in Toronto, claimed that tender points were objective findings, precisely located at 14 common sites [10]. Suddenly, the concept of reliable and objective tender points became central to the diagnosis of fibrositis, which, after changing its name to primary fibromyalgia, became widely accepted by rheumatologists throughout the world.

Clinical features of fibromyalgia

Primary fibromyalgia is currently defined as a disorder characterized by chronic, generalized musculoskeletal pain, the presence of 7 or more (out of 14) tender points, and the absence of a known specific cause. The onset of symptoms is usually between the ages of 40 and 60, although onset in childhood has been reported. At least 85% of patients are female. The majority of patients report additional symptoms of disturbed sleep and excessive fatigue, and many also complain of headaches, irritable bowel symptoms, anxiety and depression. It is axiomatic that there are no objective abnormalities on physical examination or in standard laboratory tests. The popularity of this diagnosis is attested by a prevalence of between 6% and 15% among patients in university rheumatology clinics [11].

From the first, authorities disagreed about whether palpable nodules were a consistent feature of tender points. At any rate, the muscle spasm theory was disproven by multiple studies showing that tender points are electrically silent by needle EMG [12]. Most of the 14 classical tender points are, in fact, located near muscle insertions or joints, not in muscle bellies. Although pain may be relieved by injection of a tender point with a local anesthetic, injection of normal saline – and even dry needling – are equally efficacious [3]. The specificity of the tender points is further called into question by the fact that most patients exhibit diffuse skin-fold tenderness [11]. In addition, many patients with generalized myalgia lack tender points, but are otherwise indistinguishable from those with "typical" fibromyalgia. In recent writings, one hears less about the central role of tender points in the diagnosis of fibromyalgia.

Once the primacy of tender points has been abandoned, fibromyalgia becomes a mere synonym for unexplained, generalized soft-tissue pain. Of course, there are many well-accepted disorders that include this symptom. As listed in Table 1, these diagnoses depend on objective clinical and laboratory findings. In contrast, laboratory investigations have repeatedly failed to provide consistent evidence for the existence of a physical disease in fibromyalgia.

Table 1. Recognized diseases with generalized muscle pain

Disease	Objective features
Hypothyroidism	CPK; thyroid function tests
Polymyalgia rheumatica	Sedimentation rate
Polymyositis	Weakness; CPK; EMG and muscle biopsy
McArdle's disease	Exertional cramp; CPK; myoglobinuria
Mitochondrial myopathy	Weakness; lactic acidosis; muscle biopsy
Osteomalacia, hyperparathyroidism	Bone pain and tenderness, bone X-rays, calcium metabolism
Parkinsonism	Neurological examination

Laboratory investigations

Routine clinical laboratory studies are normal, including sedimentation rate, serum creatine kinase levels, and electromyography. As mentioned earlier, muscle from fibromyalgia patients has been examined pathologically since the nineteenth century, and many different abnormalities have been reported, none of which were consistent. Even in the modern era of histochemistry and electron microscopy, abnormalities continue to be reported, but when controlled, blinded studies are performed, the abnormalities usually vanish. Abnormalities of muscle energy metabolites have also been found, but magnetic resonance spectroscopy of muscle was normal. Exercise physiology testing gave normal results when compared to sedentary control patients. All of these studies have been critically reviewed by Simms [12]. As Osler remarked, with considerable foresight, "It is by no means certain that the muscular tissues are the seat of the disease" [13].

A cerebral origin of symptoms has also been investigated. Moldofsky [14] thought he had discovered a non-REM sleep disorder specific to fibromyalgia, and he postulated that sleep deprivation was the cause of the muscle pain. Those claims were subsequently withdrawn when appropriate control patients were studied. More fruitful results have been obtained, however, from studies of pain physiology. Several research groups have found that fibromyalgia patients have diffusely low pain thresholds to pressure, cold, and heat when compared to normal controls [15]. Comparison with rheumatoid arthritis patients shows that fibromyalgics have a reduced pain threshold for pressure, and a reduced tolerance of both pressure and noise [16]. In fibromyalgia, pressure-pain thresholds increase normally during vibratory stimulation, which probably inhibits pain transmission segmentally in the dorsal horn of the spinal cord. However, pressure-pain thresholds do not increase following "heterotopic noxious conditioning stimulation" (HNCS) produced by tourniquet exercise [17]. HNCS is thought to act through pathways descending from supraspinal structures, which inhibit wide-dynamic-range neurons in the dorsal horn. Thus, there is evidence of altered cerebral processing of pain information in fibromyalgia, with lower pain thresholds and reduced central inhibition.

Fibromyalgia and depression

Almost from the beginning, some physicians have postulated a psychiatric origin for fibromyalgia. Several studies do, in fact, suggest a significant link to depression. The most impressive data are those of Goldenberg [18], which showed a lifetime history of major depression in 71% of fibromyalgia patients, five times the incidence of 14% found in rheumatoid arthritis patients. Very similar percentages (without arthritis controls) were reported by Buchwald [19]. Furthermore, 10% of first-degree relatives of fibromyalgia patients have affective disorders, compared with 3% of relatives in rheumatoid arthritis [18]. Although these data are compelling. the most intriguing finding was that only one-fourth of fibromyalgia patients satisfied the diagnostic criteria for major depression *at the time of evaluation*

[18, 19]. Indeed, all authors agree that antidepressant therapy is not very helpful in fibromyalgia. From these data, Hudson and Pope [20] have concluded that, although fibromyalgia and depression are closely related, depression does not 'cause' fibromyalgia, nor vice versa. The two disorders may be caused by some underlying factor, as yet unidentified.

Symptom or disease?

From the foregoing discussion we can draw the following conclusions about fibromyalgia:
1) there is no convincing evidence for a primary disorder of muscle or connective tissue;
2) there is evidence of perceptual amplification of noxious stimuli;
3) there is a link with major depression;
4) hence, the symptom of muscle pain is probably of cerebral origin.
Elsewhere, I have advanced similar arguments about the cerebral origin of the symptom of asthenia in the chronic fatigue syndrome [21]. The latter syndrome overlaps considerably with fibromyalgia (Table 2).

Table 2. Frequency of various symptoms in patients with chronic fatigue syndrome (CFS), fibromyalgia (FM), or both (CFS + FM)

Symptoms	CFS	FM n (%)	CFS + FM
Muscle "weakness"	211 (89)	52 (85)	109 (96)
Myalgia	234 (99)	61 (100)	114 (100)
Postexertional fatigue	153 (65)	21 (34)	86 (75)
Headaches	206 (87)	36 (59)	102 (90)
Migratory arthralgia	161 (68)	31 (51)	59 (52)
Cognitive difficulty	234 (99)	58 (95)	114 (100)
Sleep disturbance	230 (98)	55 (90)	109 (96)
Diarrhea	90 (38)	20 (33)	55 (48)
Spastic colon	36 (30)1	10 (26)	17 (28)
Bedridden or shut-in	42 (18)	14 (23)	35 (31)

Adapted from [19]

In fact, chronic fatigue syndrome and fibromyalgia are only two of the many diagnoses that have been mined from the lode of symptoms which the neurologist George Beard, in 1870, originally grouped under the name of neurasthenia [21, 22]. Patients with primarily cardiac symptoms (chest pain, shortness of breath, tachycardia, and exercise intolerance) have been reported as effort syndrome, neurocirculatory asthenia, or mitral valve prolapse. Patients with prominent hyperventilation symptoms were said to have spasmophilia or intracellular magnesium deficiency. Predominant psychiatric and pseudoneurological symptoms were called benign myalgic encephalomyelitis. All of these disorders share the same set of symptoms; it is a question of which symptom the physician chooses to consider the primary one.

Then is fibromyalgia a disease? The answer depends, in part, on what we mean by a disease. Some neurological conditions have no physical signs or laboratory findings, yet are well-accepted as diseases. These include migraine, restless legs syndrome, writer's cramp, and Tourette syndrome. All of these conditions have, as recently as forty years ago, been considered psychogenic by some physicians. What led to the universal acceptance of these conditions as "organic"? I think the most important reasons were 1) the distinctive and stereotyped character of the symptoms; 2) the fact that most patients appeared psychiatrically normal, and 3) the absence of multiple other unexplained symptoms. Fibromyalgia does not meet these criteria.

But if fibromyalgia is a psychiatric disorder, so is major depression. Most physicians now accept that depression has an important "biological" component, manifested by altered neurotransmitter and neuroendocrine function, and influenced to varying degrees by genetic predisposition. Thus, if fibromyalgia is a kind of depression-equivalent, there could be some (undiscovered) neurochemical alteration that accounts for the low pain threshold and the multiple somatic symptoms. Many would argue that such a putative alteration of brain chemistry would also qualify as a disease. On the other hand, patients with fibromyalgia often have clinical characteristics closer to somatization disorder than to depression. But even somatization disorder, which has less of a "biological" cachet than depression, appears to have genetic components [23], and it certainly behaves like a chronic, incurable disease.

When patients ask for an explanation of their symptoms, what should we say? Since we do not actually know the correct cause, it is reasonable to tell them what we do know. Barsky [24] recommends telling a somatizing patient that "the problem lies in the nervous system's processing of bodily sensations, and that the patient has an extraordinarily sensitive nervous system that amplifies benign symptoms that others experience as less intense". Re-educating patients along these lines is the basis of cognitive-behavioral therapy. Such treatment has not been uniformly successful [25, 26], although it has the virtue of intellectual honesty. But most patients with fibromyalgia, like those with chronic fatigue syndrome, do not welcome being told that their symptoms are "psychological," or that they do not have a physical illness. Commonly, they seek out physicians who will give them a physical diagnosis, often of dubious validity.

Finally, there are important legal implications about the status of fibromyalgia as a disease. In 1995, a review of 1,668 fibromyalgia patients from 7 United States medical institutions revealed that 25% had received some type of disability payment. The estimated annual cost of disability payments in the United States alone is $8 billion [27]. What should a physician write when asked to fill out a disability application? In a time of shrinking social welfare funds, should patients with fibromyalgia, a condition with no objective, verifiable diagnostic criteria, be eligible for medical disability pensions?

References

1. Mills KR, Edwards RHT (1983) Investigative strategies for muscle pain. J Neurol Sci 58:73-88
2. Serratrice G et al (1980) À propos de 210 cas de myalgies diffuses. Sem Hôp (Paris) 56:1241-1244
3. Simons DG (1975) Muscle pain syndromes. Am J Phys Med 54:289-311; and (1976) 55:15-42
4. Gowers, WR (1904) Lumbago: its lessons and analogues. Brit Med J 1:117-121
5. Stockman R (1904) The causes, pathology and treatment of chronic rheumatism. Edinburgh Med J 15:107-116, 223-235
6. Llewellyn LJ, Jones AB (1915) Fibrositis. Heinemann, London
7. Hench PS (1936) The problem of rheumatism and arthritis (third rheumatism review). Ann Intern Med 10:880
8. Ellman P, Shaw D (1950) The "chronic rheumatic" and his pains. Psychosomatic aspects of chronic non-articular rheumatism. Ann Rheum Dis 9:341-357
9. Travell J, Rinzler SH (1952) The myofascial genesis of pain. Postgrad Med 11:425-434
10. Smythe H (1986) Tender points: evolution of concepts of the fibrositis/fibromyalgia syndrome. Am J Med 81 [Suppl 3A]:2-6
11. Wolfe F (1986) The clinical syndrome of fibrositis. Am J Med 81 [Suppl 3A]:7-14
12. Simms RW (1996) Is there muscle pathology in fibromyalgia syndrome? Rheum Dis Clin N Am 22:245-266
13. Osler W (1909) The principles and practice of medicine. 7th edn, D. Appleton and Co, New York, London p 396
14. Moldofsky H et al (1975) Musculoskeletal symptoms and non-REM sleep disturbance in patients with "fibrositis syndrome" and healthy subjects. Psychosom Med 37:341-351
15. Kosek E, Ekholm J, Hansson P (1996) Sensory dysfunction in fibromyalgia patients with implications for pathogenetic mechanisms. Pain 68:375-83
16. McDermid AJ, Rollman GB, McCain GA (1996) Generalized hypervigilance in fibromyalgia: evidence of perceptual amplification. Pain 66:133-44
17. Kosek E, Hansson P (1997) Modulatory influence on somatosensory perception from vibration and heterotopic noxious conditioning stimulation (HNCS) in fibromyalgia patients and healthy subjects. Pain 70:41-51
18. Goldenberg DL (1986) Psychologic studies in fibrositis. Am J Med 81 [Suppl 3A]:67-70
19. Buchwald D (1996) Fibromyalgia and chronic fatigue syndrome. Similarities and differences. Rheum Dis Clin N Am 22:219-243
20. Hudson JI, Pope JG Jr (1996) The relation between fibromyalgia and major depressive disoder. Rheum Dis Clin N Am 22:285-303
21. Layzer RB (1989) Muscle fatigue. In: Serratrice G et al (eds) Myélopathies, Neuropathies et Myopathies: Acquisitions Récentes. Expansion Scientifique Française, Paris, pp 388-394

22. Wessely S (1990) 'Old wine in new bottles': neurasthenia and 'ME'. Psychol Med 20:35-53
23. Goodwin DW, Guze SB (1989) Psychiatric Diagnosis. 4th edn, Oxford University Press, New York, p 118
24. Barsky AJ (1997) Clinical Crossroads. A 37-year-old man with multiple somatic complaints. JAMA 278:673-679
25. Burckhardt CS, Mannerkorpi K, Hedenberg L, Bjelle A (1994) A randomized, controlled clinical trial of education and physical training for women with fibromyalgia. J Rheumatol 21:714-720
26. White KP, Nielson WR (1995) Cognitive behavioral treatment of fibromyalgia syndrome: a followup assessment. J Rheumatol 22:717-21
27. Wolfe F, Potter J (1996) Fibromyalgia and work disability. Is fibromyalgia a disabling disorder? Rheum Dis Clin N Am 22:369-91

Is Chronic Fatigue Syndrome a Muscle Disorder?

W.M.H. Behan

Chronic fatigue syndrome (CFS) is a clinically defined disorder characterized by severe disabling fatigue made worse by exertion and accompanied by a variety of other symptoms, including lapses in memory and concentration and sleep disturbances [1, 2]. Its major symptom, fatigue, is common to many other diseases and there is no objective diagnostic test so that it has proved extremely difficult to establish clinical guidelines. The Working Case Definition produced by the Centers for Disease Control in 1988, followed by the report of a Consensus Conference from the United Kingdom were helpful but it became obvious that certain major problems still needed to be addressed, i.e. fatigue had to be defined more rigorously and the relationship had to be established between CFS and psychiatric disorders, especially depression, and certain poorly defined syndromes such as fibromyalgia and neurasthenia [2]. The definition of fatigue was especially critical since it is such a common symptom in medical practice: indeed 24% of the adult population of the USA are said to have had episodes lasting for two weeks or longer, unexplained in the majority [2].

A revised scheme, a conceptual framework, has therefore been put forward (Table 1): this is widely used and has been seminal in diagnosis [2]. The distinction between patients with chronic fatigue, the majority of whom will turn out to have a psychiatric illness, and a clearly defined subgroup with CFS, has been elucidated. Detailed studies have illustrated the fundamental point that most patients with 'fatigue' do not have the specific disorder, CFS, although there is a distinct subset who do [2, 3]. There is still disagreement between those who maintain that, since the majority of cases have a psychiatric disorder, CFS is a psychiatric illness, and those who hold the opposite view maintaining that there is no psychiatric component. This appears as controversy over the use of a broad diagnostic definition or of a restrictive approach in which all the criteria must be fulfilled [4]. The important feature of the latest guidelines, however, is that the conceptual framework proposed allows for both views, provided the one held is stated clearly [2]. Thus, this scheme can be expected to improve the standard of research into pathogenesis and therapy.

Prevalence rates vary, being estimated as 0.9 to 1% in the USA using restrictive criteria and up to 2.6% in the UK by psychiatrists using a broader definition [4]. We exclude all patients with depression and other psychiatric problems in our studies, since we consider that only then does research on CFS become meaningful.

Table 1. Definition of CFS

- The presence of clinically evaluated, unexplained, persistent **chronic fatigue**, lasting six months or more, of new onset, not related to exertion or cured by rest and resulting in a substantial reduction in activity.
- The concurrent occurrence of four or more symptoms
  - post exertional myalgia
  - impairment in short-term memory or concentration
  - sleep disturbances
  - sore throat, muscle or joint pain
  - headaches of a new type or severity
- The following are exclusion criteria
  - any active medical condition which can produce chronic fatigue
  - any post or current major depressive illness, bipolar affective disorder or schizophrenia
  - any evidence of alcohol or substance abuse for two years before onset or afterwards
  - severe obesity

Muscle findings

Histological studies

There is such a discrepancy between the severity of the patients' complaints of muscle fatigue and myalgia and the lack of specific histological features that there has inevitably been suspicion that this is a psychiatric illness. CFS, however, is not the only muscle disorder in which there is no morphological evidence of muscle damage: the original scenario of myasthenia gravis was much the same.

The arguments about the histological features of muscle in CFS can be reduced to attempts to answer two questions: first, are there any abnormalities and secondly, if there are, are they simply the result of inactivity?

The first question has been addressed in several studies and the consensus is that a variety of findings are present, all non-specific and none associated consistently [5-14]. The features described include changes in fibre type predominance, focal or diffuse atrophy, occasional fibre necrosis, glycogen depletion, abnormal oxidative enzyme staining and evidence of increased regeneration, but most workers agree with Edwards et al. [11] who reported the largest study, that less than 1% of fibres are affected. In addition, exactly the same morphological changes - focal necrosis, regeneration, increased or decreased lipid and glycogen content, lymphocytic infiltrates have been identified in control and patient biopsies except to a different degree - 32% of controls and 81% of patients [11].

As regards atrophy, mean fibre areas are within the normal range: only two of 108 cases showed type 2 fibre atrophy while three others showed hypertrophy of type 2 and/or type 1 fibres, in the largest study [11]. Scattered atrophic fibres have been reported in most studies [5, 6, 8, 14]. No consistent fibre type prevalence has been identified [11]. Approximately the same incidence of myoadenylate deaminase deficiency is present (3 of 108) as in the normal population, where 1-2% of muscle biopsies are affected [15]. We identified eight cases in the 400 biopsies of CFS which we have examined.

No specific mitochondrial changes were seen by Edwards and coworkers, [11] although 4% had increased mitochondrial numbers, confirmed on ultrastructure and accompanied in 2% by partial cytochrome oxidase deficiency. Another group using functional mitochondrial assays reported no changes of significance [9].

We looked at muscle ultrastructure in biopsies from 50 patients with CFS and detected increased numbers of subsarcolemmal mitochondria in the majority, with frequent branching and fusion of mitochondrial cristae leading to a 'honeycombed' or compartmentalised appearance [7]. Morphological changes in mitochondria are not specific and may accompany several neuromuscular disorders in which there is no apparent metabolic defect: nonetheless, florid proliferation and structural abnormalities tend to be associated with disorders of the respiratory chain or of oxidative phosphorylation coupling [16]. The appearance may reflect high metabolic activity or a disturbance in the usual sequence of metabolic steps, leading to accumulation of a metabolite and attempts at compensation. In our study we saw the same changes, but very rarely and to a much lesser degree, in the control muscle samples taken from the operative site of older men and women undergoing surgery for benign or malignant breast tumours, hernias or vascular problems [7].

The CFS group was not only younger than the control group but also included, as is typical, cases who were very fit before the onset of the disorder and this may be relevant since there can be striking mitochondrial proliferation and enlargement in young athletes. Other groups have not confirmed our findings [11, 17, 18]: some of this may be due to patient selection and some to the same phenomenon as the histological findings, i.e. non-specific features are present more consistently in patients than in controls.

There is certainly other evidence of mitochondrial dysfunction in CFS namely, decreased levels of serum acylcarnitine, an increased incidence in the common $mtDNA^{4977}$ deletion and unusual mtDNA deletions [19, 20, 21]. The latest reports of functional muscle studies (see below) are also in keeping.

From all these reports it may be concluded that there are no diagnostic muscle biopsy features in CFS, but this still leaves the second question: is there evidence of deconditioning as the result of inactivity? This has been answered recently in a detailed study of 105 biopsies, examined by a semi-automated image analysis system in order to establish the proportions of type 1 to type 2 fibres and the degree of muscle atrophy and hypertrophy [17]. Evidence of a shift to type 2 fibre predominance was sought, to indicate deconditioning as the result of inactivity.

The results revealed no such evidence of inactivity: indeed type 1 fibre predominance (23%) was more common than type 2 (3%). Fibre atrophy (of type 1 and/or type 2) was detected in only 10.4% of cases, fibre hypertrophy being more common and present in 56%. Interestingly, some of a subgroup of cases with abnormal lactate responses to exercise in the subanaerobic threshold test (SATET) showed a relative deficiency in mitochondria-rich type 1 fibres, in keeping with reduced oxidative metabolism [17]. Inconsistent and non-specific abnormalities were present, as in all studies of CFS.

The authors suggested their findings had a biological significance [17]: type 2 fibre predominance is known to occur where static strength is preserved at the expense of endurance. Thus, subjects with type 2 fibre predominance tend to have poorer exercise performance, probably due to the relatively inefficient production of ATP in these fibres compared to the type 1 mitochondria-rich fibres. For instance, studies of patients with idiopathic exertional myalgia and fatigue [22] and military recruits who have heatstroke complicated by rhabdomyolysis [23] show a significantly increased prevalence of type 2 fibres.

Therefore, the numerous morphological studies of muscle in CFS can be summarised as revealing non-specific abnormalities only and, most importantly, no evidence of deconditioning or disuse atrophy.

Functional studies

Is CFS a muscle disease? Functional studies indicate that it is indeed: abnormalities can be detected, providing evidence of metabolic lesions and interesting in that they may provide pathogenetic clues.

The functional exercise studies which were originally carried out gave contradictory results: some providing support for a central origin for the fatigue, i.e. a lack of motivation [24-29] and others for a peripheral origin, i.e. a metabolic muscle cause [30]. There was a great deal of controversy before the advent of phosphorus ^{31}P Nuclear Magnetic Resonance spectroscopy (^{31}P NMR) changed the picture completely because of its ability to provide continuous *in vivo* monitoring of intracellular energy metabolism. Studies using ^{31}P NMR have now confirmed that there is a significant reduction in exercise capacity in CFS [31-35]. The patients reach exhaustion more rapidly than normal subjects, in keeping with an abnormality in oxidative metabolism and a resultant acceleration of glycolysis in the working skeletal muscle. There is significantly decreased muscle aerobic metabolism, consistent with a subtle mitochondrial abnormality, as measured by estimating recovery in phosphocreatine (PCr) levels [34].

It is worth examining in detail the recovery phase after exercise, since patients complain consistently that this is prolonged. My colleagues and I [36] carried out a fatiguing test in which the objective was to concentrate on this period. The quadriceps muscle groups of 10 patients and 10 controls were tested in a 50% duty cycle of 18 maximum voluntary cotractions (MVCs), with the force generated by each contraction recorded using a KinCom dynamometer. Two recovery periods were studied: the first lasting 200 minutes immediately after the test and the second, at 24 hours post-exercise.

Recovery was prolonged in the patient group with a statistically significant difference compared to initial MVCs being evident in the first recovery period ($p < 0.01$) and more dramatically increased at 24 hours ($p < 0.001$). In contrast, the control group achieved MVCs which were not significantly different from their initial values, at both periods. This finding supports the clinical complaint of delayed recovery after exercise in patients with CFS.

The test we used employed repetitive isometric contractions and has a high metabolic cost with a continuously increasing energy demand and a metabolic effect afterwards. The relative force produced by the patient group was less than in the control, but the relative decrease in the two groups was identical, indicating that there is normal muscle endurance on exercise, as documented elsewhere. Others have not examined the post-exercise period in such detail although interestingly enough, analysis of their data suggests a similar result [26]. A different exercise regime was used by Gibson et al. [24], making it difficult to compare results.

In vitro studies

My colleagues and I have also recently completed another study of interest: since a central effect has been implicated so often in the muscle studies, we decided to obviate this completely by carrying out an *in vitro* analysis of myoblast cultures obtained from needle biopsies of patients, the first study in which continuous, long-term assessment of muscle respiration has been made. We investigated aerobic metabolism in 16 cases of CFS compared to 10 healthy controls [37].

The respiratory chain, lactic and pyruvate production, cytochrome oxidase and lactic dehydrogenase activities were examined, using well-established micro-methods [38]. Total mitochondrial DNA volume was also measured, two mitochondrial DNA re-arrangements and one point mutation sought. The most informative test of those carried out proved to be assessment of the cells' redox state, derived by comparing the steady-state levels of lactate and pyruvate (lactate/pyruvate, L/P ratio) in the spent medium.

Our results [37] showed that the L/P ratios were spread significantly more widely ($p = 0.011$) in the samples from patients with CFS compared to controls, using F variance ratio or F-test. There was no evidence that the two populations were distributed around different means or different medians and therefore the patient group was the more heterogeneous. The results are shown in Fig. 1.

Muscle cells from two patients had increased L/P ratios, suggestive of mild mitochondrial dysfunction, while eight had decreased ratios, raising the possibility of pyruvate dehydrogenase (PDH) deficiency. The remaining six myoblast cultures had a ratio within the same range as the controls. Respiratory chain complexes were normal.

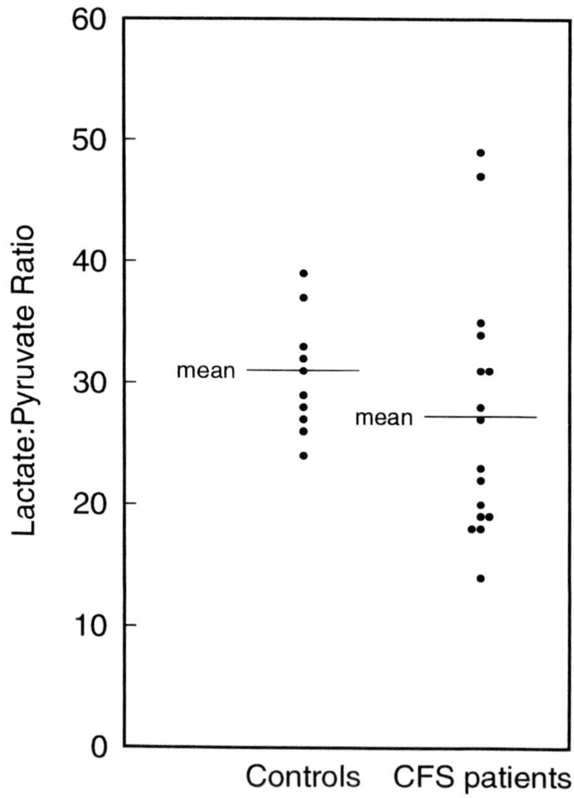

Fig. 1. Lactate/Pyruvate ratio in myoblast cultures from patients with CFS compared to controls

The L/P ratio in cultured cells has been used extensively in studies of infants with severe metabolic defects [38], but clearly we would not have predicted L/P ratios over 100, similar to such cases, in the patients with CFS.

In the two cases with a small but obvious increase in the L/P ratio, an increase in lactate production was demonstrated. In the sub-group with low L/P ratios, there was no difference between lactacte production in the patients' muscle cells compared to controls but there was a difference in the level of the pyruvate produced, just failing to achieve significance ($p = 0.053$ by the t-test). Since LDH activity appeared normal, mild PDH deficiency is the most plausible explanation for the decreased L/P ratios in this group. Decreased flux through glycolysis appears unlikely as it would be expected to lead to a low level of pyruvate, and possibly also to low levels of lactate.

Finally, in our study, quantitative analysis of mitochondrial DNA levels demonstrated a small but insignificant increase in patients with CFS, compared to healthy controls while none of the patients or disease controls had deletions in $mtDNA^{4977}$ or $mtDNA^{7436}$ by PCR, or any other large scale rearrangements by Southern blotting. All were negative for the $mtDNA^{3243}$ mutation associated with MELAS. Mild non-specific ultrastructural abnormalities, rare paracrystalline inclusions, numerous pleomorphic mitochondria with proliferation of cristae and increased lipid were present in muscle biopsies from nine of the 16 cases, including in seven of those with low L/P ratios [37].

Viruses in muscle

Since CFS is a disorder in which approximately one-third of the patients describe an acute onset related to a febrile illness, a viral aetiology has always been postulated [1]. There are compelling historical accounts of epidemics [1] but recent work by the CDC has not revealed recent evidence of epidemic spread, and it is now considered that the disorder mainly has a sporadic pattern [2] with rare epidemic outbreaks. This may, in fact, have always been the case since individual patients with CFS were diagnosed both before and after the well known Los Angeles and Royal Free epidemics [1].

The most attractive hypothesis which implicates viruses in the disorder has been to suggest that there is a persistent virus present in the affected muscle [1, 5]. Oldstone [39] has demonstrated conclusively in animal models that such an agent may linger in the central nervous system or other tissues, cause profound changes by interfering with the function of differentiated cells and yet produce no histological evidence, no inflammation or cell necrosis, to indicate infection [39].

At least 11 different RNA and DNA viruses have been associated with CFS, however, and none has been confirmed as persistent. The human herpes viruses, Epstein-Barr, human herpes virus type 6 (HHV-6) and cytomegalovirus have been particularly implicated, but it appears likely that they are reactivated in the host as the result, not the cause, of altered immune function [40].

The role of enteroviruses, especially the ubiquitous and myotropic coxsackie B, has attracted a great deal of attention in CFS. The first molecular studies were positive for enterovirus in muscle [5] and indeed the presence of a replication-deficient mutant, in keeping with a persistent agent was reported. A larger study using more rigorous techniques, however, revealed 26.4% positivity for enteroviral genome in muscle from the patient group compared to 19.8% in the controls, a difference which was not significant and militates against a role for these viruses [41].

Great interest was aroused by a report from a group at the Wistar Institute in Philadelphia that retroviral sequences (HTLV 1/11) were present in the peripheral blood cells of the majority of cases of CFS, but we found no evidence of human T lymphotropic-like or foamy (spuma) viruses in muscle biopsies or peripheral blood from 30 well characterised patients with CFS, using more detailed PCR primer assays to amplify *gag*, *pol*, *env* and *tax* regions of HTLV 1/11 [42]. What we did find was an endogenous *gag* band in tissue from patients and controls and we consider that this accounted for the positive result reported [42].

Another virus sought has been the human parvovirus B19, known to persist in human tissues and associated with fibromyalgia - a disorder considered by many authorities to be identical to CFS. However, our studies of muscle biopsies revealed no increased incidence of this virus [43].

The latest virus to attract attention in CFS is the Borna disease virus (BDV) which has been associated with affective disorders and neurobehavioural syndromes in man and, recently, with CFS. We examined peripheral blood leucocytes, serum and cerebrospinal fluid from 25 cases of CFS together with brain tissue from one case, compared to controls [44]. Reverse transcriptase polymerase chain reaction (RT-PCR) was employed on RNA from the samples, followed by Southern blot analysis and hybridisation with an end-labelled oligonucleotide probe. Two serum samples and one cerebrospinal fluid from patients with CFS were positive for BDV while all control samples were negative. Thus, no significant association between BDV and CFS was demonstrable [44].

At the present time, therefore, there is no evidence to implicate a persistent virus in CFS.

Other possible factors

A novel possibility raised recently is that CFS is an ion-channel disease, a channelopathy [45-47]. These disorders affect a class of proteins which generate "the electrical signals passing through the thinking brain, the beating heart and the contracting muscle" [48]. They can be inherited or acquired and were first recognised in the familial periodic paralyses and non-dystrophic myotonias (e.g. Thomsen's disease). Typically, paroxysmal attacks occur, precipitated by physiological stress. The clinical similarities between CFS and Syndrome X - a cardiological disorder in which angina-type pain and abnormal thallium cardiac SPECT scans are found in the absence of ischaemic heart disease - first suggested that abnormal ion channels might be present since excessive potassium efflux through cardiac cell membranes has been postulated to play a role in the latter [47]. Such an abnormality would also have an effect on resting energy expenditure (REE) since 30% of REE is normally used to maintain sodium-potassium cell concentration gradients. A small study recently reported by our group documents an increase in REE in 50% of the cases of CFS examined, compared to 11 age, height, sex and weight-matched controls [47].

Acquired channelopathies may be caused by viruses or toxins, e.g. picornaviruses may attach to receptors in ion-channels, maintaining them in a permanently open state [45-49] while a fish poison, ciguatera toxin, produces its effects by binding to, and keeping open, voltage-gated sodium channels [50]. It is of great interest therefore that poisoning with this toxin may result in a syndrome clinically identical to CFS [50]. This potent, tasteless and heat-stable agent is passed up the piscine food chain to reach the 'gourmet' fish eaten by humans and usually produces an acute syndrome, which resolves spontaneously. Just under 2% of cases, however, develop CFS which may last up to 3 years before remission [50]. Therapy for these channelopathies is being developed: some membrane-stabilising drugs are already available and specific channel-modulating agents are being developed: thus, therapy for CFS may at last be near.

Conclusions

Review of the data on muscle in CFS reveals that this is likely to be a heterogeneous syndrome. There are no consistent morphological muscle abnormalities nor is there any evidence of disuse atrophy. Non-specific features may, however, be more common than in normal muscle. Both *in vivo* and *in vitro* evidence of dysfunctional oxidative metabolism can be demonstrated. With regard to pathogenesis, it seems unlikely that a persistent virus is the cause since herpesviruses, enteroviruses, retroviruses, parvovirus and Borna disease virus have all been excluded. A novel hypothesis under consideration is that CFS is a muscle channelopathy.

Acknowledgements

The work of myself and my colleagues in the Chronic Fatigue Syndrome Group at Glasgow University is supported by the Barclay Trust, the Persistent Virus Disease Research

Foundation, the Myalgic Encephalomyelitis Association, the Cunningham Trust, the Linbury Trust, the Wekerley Trust and Action for ME.

References

1. Behan PO, Behan WMH (1988) Postviral fatigue syndrome. CRC Crit Rev Neurobiol 4:157-178
2. Fukuda K, Straus SE, Hickie I et al (1994) The chronic fatigue syndrome: a comprehensive approach to its definition and study. Ann Intern Med 121:953-959
3. Walker EA, Katon WJ, Jemelka RP (1993) Psychiatric disorders and medical care utilization among people in the general population who report fatigue. J Gen Intern Med 206:929-934
4. Editorial (1996) The Lancet 348:971
5. Archard LC, Bowles NE, Behan PO et al (1988) Postviral fatigue syndrome: persistence of enteroviral RNA in muscle and elevated creatine kinase. J Soc Med 81:326-329
6. Behan PO, Behan WMH, Bell EJ (1985) The postviral fatigue syndrome – an analysis of the findings in 50 cases. Journal of Infection 10:211
7. Behan WMH, More IAR, Behan PO (1991) Mitochondrial abnormalities in the postviral fatigue syndrome. Acta Neuropathol 83:61-65
8. Byrne E, Trounce I, Dennett X (1985) Chronic relapsing myalgia (post viral ?): clinical, histological and biochemical studies. Aust NZ J Med 15:305-308
9. Byrne E, Trounce I (1987) Chronic fatigue and myalgia syndrome: mitochondrial and glycolytic studies in skeletal muscle. J Neurol Neurosurg Psychiatry 50:743-746
10. Connolly S, Smith D, Doyle D et al (1993) Chronic fatigue: electromyographic and neuropathological evaluation. J Neurol 240:435-438
11. Edwards R, Gibson H, Clague J et al (1993) Muscle physiology and histopathology in chronic fatigue syndrome. In: Kleinman A, Straus S (eds) Chronic fatigue syndrome. John Wiley, Chichester, pp 101-131
12. Grau JM, Casademont J, Pedrol E et al (1992) Chronic fatigue syndrome: studies on skeletal muscle. Clin Neuropathol 11:329-332
13. Karpati G, Carpenter S, Weller B et al (1990) Muscle biopsy experience in the chronic fatigue syndrome (abstract). J Neurol Sci 98 (Suppl):96
14. Preedy VR, Smith DG, Salisbury JR et al (1993) Biochemical and muscle studies in patients with acute onset post-viral fatigue syndrome. J Clin Pathol 46:722-726
15. Fishbein WN (1985) Myoadenylate deaminase deficiency. Inherited and acquired forms. Biochem Med 33:158-169
16. Di Mauro S, Bonilla E, Zeviani M et al (1985) Mitochondrial myopathies. Ann Neurol 17:521-538
17. Lane RJM, Barrett MC, Woodrow D et al (1998) Muscle fibre characteristics and lactate responses to exercise in chronic fatigue syndrome. J Neurol Neurosurg Psychiatry 64:362-367
18. Pliophys AV, Pliophys S (1995) Electron-microscopic investigation of muscle mitochondria in chronic fatigue syndrome. Neuropsychobiology 32:175-181
19. Behan WMH, Gow JW, Simpson K et al (1994) Mitochondrial findings in the chronic fatigue syndrome. J Pathol 173:S17
20. Kuratsune H, Yamaguti M, Misaki H, Tagawa S, Kitani T (1994) Acylcarnitine deficiency in chronic fatigue syndrome. Clin Inf Dis 18 (Suppl 1):S62-67
21. Zhang C, Baumer A, Mackay IR, Linnane AW, Nagley P (1995) Unusual pattern of mitochondrial DNA deletions in skeletal muscle of an adult human with chronic fatigue syndrome. Hum Mol Genet 4:751-754
22. Telerman-Topet N, Bacq M, Khoubesserian P et al (1985) Type 2 fibre predominance in muscle cramp and exertional myalgia. Muscle and Nerve 8:563-567

23. Hsu Y-D, Lee W-H, Chang M-K et al (1997) Blood lactate threshold and type II fibre predominance in patients with exertional heatstroke. J Neurol Neurosurg Psychiatry 62:182-187
24. Gibson H, Carroll N, Clague JE, Edwards RHT (1993) Exercise performance and fatiguability in patients with chronic fatigue syndrome. J Neurol Neurosurg Psychiatry 56:993-998
25. Kent-Braun JA, Sharma KR, Weiner MW, Massie B, Miller RG (1993) Central basis of muscle fatigue in chronic fatigue syndrome. Neurology 43:125-131
26. Lloyd AR, Gandevia SC, Hales JP (1991) Muscle performance, voluntary activation, twitch properties and perceived effort in normal subjects and patients with the chronic fatigue syndrome. Brain 114:85-98
27. Lloyd AR, Hales JP, Gandevia SC (1988) Muscle strength, endurance and recovery in the post-infection fatigue syndrome. J Neurol Neurosurg Psychiatry 51:1316-1322
28. Riley MS, O'Brien CJ, McCluskey DR, Bell NP, Nicholls DP (1990) Aerobic work capacity in patients with chronic fatigue syndrome. Br Med J 301:953-956
29. Stokes MJ, Cooper RG, Edwards RHT (1988) Normal muscle strength and fatiguability in patients with effort syndromes. Br Med J 297:1014-1017
30. Lane RJM, Burgess AP, Flint J, Riccio M, Archard LC (1995) Exercise responses and psychiatric disorder in chronic fatigue syndrome. Br Med J 311:544-545
31. Arnold DL, Bore PJ, Radda GK, Styles P, Taylor DJ (1984) Excessive intracellular acidosis of skeletal muscles in exercise in a patient with a postviral exhaustion fatigue syndrome. Lancet I 1367-1369
32. Arnold DL, Bore PJ, Radda GK, Styles P, Taylor DJ (1984) Excessive intracellular acidosis of skeletal muscle on exercise in the post-viral exhaustion/fatigue syndrome: a ^{31}P NMR study. Proc. 3rd Annual Meeting of the Society for Magnetic Resonance in Medicine, New York, pp 12-13
33. Barnes PRJ, Taylor DJ, Kemp GJ, Radda GK (1993) Skeletal muscle bioenergetics in the chronic fatigue syndrome. J Neurol Neurosurg Psychiat 56:679-683
34. McCully KK, Natelson BH, Iotti S, Sisto S, Leigh Jr JS (1966) Reduced oxidative muscle metabolism in chronic fatigue syndrome. Muscle Nerve 19:621-625
35. Wong R, Lopaschuk G, Zhu G, Walker D, Catellier D, Burton D, Teo K, Collins-Nakai R, Montague T (1992) Skeletal muscle metabolism in the chronic fatigue syndrome. In vivo assessment by 31P nuclear magnetic resonance spectroscopy. Chest 102:1716-1722
36. Paul L, Wood L, Behan WMH et al (1999) Demonstration of delayed recovery from fatiguing exercise in chronic fatigue syndrome. Eur J Neurol 6:63-69
37. Behan WMH, Holt IJ, Kay DH et al (1998) In vitro study of muscle oxidative metabolism in chronic fatigue syndrome. J Chronic Fatigue Syndrome 5:32-37
38. Robinson BH, Glerum DM, Chow W, Petrova-Benedict R, Lightowlers R, Capaldi R (1990) The use of skin fibroblast cultures in the detection of respiratory chain defects in patients with lactic acidemia. Paediatr Res 28:549-555
39. Oldstone MBA (1989) Viruses can cause disease in the absence of morphological evidence of cell injury: implication for uncovering new diseases in the future. J Infect Dis 159:384-389
40. Levy JA (1994) Part III Viral studies of chronic fatigue syndrome. Clin Inf Dis 18 (Suppl 1):S117-120
41. Gow JW, Behan WMH, Simpson K et al (1994) Studies on enterovirus in patients with chronic fatigue syndrome. Clin Inf Dis 18 (Suppl 1):S126-129
42. Gow JW, Simpson K, Schliephake A et al (1992) Search for retrovirus in the chronic fatigue syndrome. J Clin Pathol 45:1058-1061
43. Kerr JR, Barrett A-M, Curran MD et al (1997) Parvovirus B19 and chronic fatigue syndrome. J Chronic Fatigue Syndrome 3:101-107
44. Gow JW, de la Torre JC, Behan WMH et al (1997) Borna disease virus in chronic fatigue syndrome. Neurological Infections and Epidemiology 2:63-66

45. Chaudhuri A, Behan WMH, Behan PO (1998) Chronic fatigue syndrome. Proc Roy Coll Physicians Edinb 28:150-163
46. Lund-Olesen LH, Lund-Olsen K (1994) The etiology and possible treatment of chronic fatigue syndrome. Med Hypotheses 43:55-58
47. Watson WS, Chaudhuri A, McCreath GT, Behan PO (1999) A possible cell membrane defect in chronic fatigue syndrome and syndrome X. Chapter 13 In: Chest Pain with Normal Coronary Angiopathy. Pathogenesis, Diagnosis & Management. Ed. By JC Kaski. Published by Kluwer Academic Publications Boston, USA in the Series Developments in Cardiovascular Medicine.
48. Ackerman MJ, Clapham DE (1997) Ion channels – basic science and clinical disease. New Eng J Med 336:1575-1586
49. Garry RF (1992) Similarities of viral proteins to toxins that interact with monovalent cation channels: molecular mimicry and a possible role in cytopathogenesis. J Cell Biochem S16c:129
50. Pearn J (1997) Chronic fatigue syndrome-chronic ciguatera poisoning as a differential diagnosis. Med J Australia 166:309-310

Exercise Intolerance and Muscle Pain in Myotonic Disorders

R.C. Griggs, V. Sansone, G. Meola, R.T. Moxley III

The cardinal features of the nondystrophic myotonias and the periodic paralyses are episodic weakness and muscle stiffness. In the past, myotonia was often defined as "painless muscle stiffness". However, the recent molecular characterization of the myotonic disorders has redefined the boundaries of each of the diseases and has also identified "new" diseases, a number of which are characterized by muscle pain as their presenting or major manifestation. In many patients this pain limits activity and produces exercise intolerance. Three disorders are particularly notable for pain: sodium channel myotonia, proximal myotonic myopathy (PROMM), and some forms of chloride channel myotonia. It is not known whether the pain arises from the muscle or whether it is even necessarily related to the myotonia. Pain in myotonic disorders is usually intermittent. Striking intensification of muscle pain by changes in physiologic variables also suggests that pain is not solely the consequence of the molecular lesion. Thus, potassium ingestion in sodium channel myotonia causes severe, life-threatening myotonia and pain. PROMM is often asymptomatic until unmasked by hypothyroidism with consequent severe and persistent pain. There have been no randomized, controlled studies of the treatment of the pain. Anecdotal experience suggests that acetazolamide and mexiletine are of benefit in sodium channel myotonia with pain and mexiletine in chloride channel myotonia. The pain of PROMM has improved following correction of hypothyroidism but has otherwise not responded to antimyotonia agents. A multicenter study is currently under development to determine the best treatment for myotonia and muscle pain.

Introduction

The clinical classification of the myotonias has been revised by the discovery of the molecular basis for most of the disorders (Table 1) [1-6]. Similarly, the nosology of the periodic paralyses has been clarified and simplified (Table 2) [3, 5, 7-11]. This paper will discuss the occurrence of pain in the myotonias and will consider two illustrative cases that raise substantive issues concerning the pathophysiology of pain attributed to myotonia and the basis for its treatment. The differential diagnosis of painful myotonia will also be considered.

Table 1. Myotonic Disorders

Pre-1992

- Myotonic dystrophy
- Congenital myotonia (Thomsen)
- Congenital myotonia (Becker)
- Congenital paramyotonia
- Hyperkalemic periodic paralysis
- Schwartz-Jampel syndrome
- Myotonia acquisita (e.g., hypothyroidism, hyperthyroidism, drug-related)
- Atypical myotonias

1988

- Myotonic dystrophy – trinucleotide repeat disease
- Proximal myotonic myopathy (PROMM)
- Chloride channel myotonias
- Chloride channel myotonia with weakness
- Sodium channel myotonias
- Schwartz-Jampel syndrome
    - chromosome 1p36 linked
    - other locus or loci
- Acquired myotonia
    - drug toxicity – colchicine, cyclosporin, clofibrate, others
    - endocrinopathy – hypothyroidism, hyperthyroidism

Myotonia has usually been characterized as "painless muscle stiffness." The recent molecular characterization of the myotonic disorders and the periodic paralyses has clarified the diagnostic categorization of these diseases and has begun to facilitate the correlation of specific molecular defects in muscle ion channels with the symptoms which result from mutations in different domains or segments of the ion channel. Thus, it is now clear that pain and exertional symptoms are features of muscle chloride [12, 13] and sodium channelopathies [14,15] as well as the recently characterized disorder, proximal myotonic myopathy (PROMM) [16, 17] (Table 3).

The molecular defects of many myotonias have been defined (Table 3). PROMM [16, 17] and Schwartz-Jampel syndrome [1, 18] may be multiple and have not been defined. The molecular defect has been defined for all periodic paralyses except Andersen's syndrome [5, 19-21].

Table 2. Primary Priodic Paralyses

Pre-1992

- Hypokalemic periodic paralysis
  - autosomal dominant
  - sporadic
- Hyperkalemic periodic paralysis
  - autosomal dominant
  - sporadic
- Myotonic periodic paralysis
- Paramyotonia
  - paralysis periodica paramyotonica
  - paramyotonia congenita
- Biphasic periodic paralysis
- Periodic paralysis with distinctive facial features, cardiac arrhythmias
- Normokalemic periodic paralysis

New Classification

- Calcium channel dihydropyridine receptor mutations – [hypokalemic]
- Sodium channel mutations – [hyperkalemic, K-sensitive]
- Chloride channel mutations [Becker's myotonia]
- Andersen's syndrome [locus, defect unknown]

Table 3. Myotonic Disorders Characterized by Pain

- Sodium channel
  - acetazolamide-responsive myotonia
  - myotonia fluctuans
  - myotonia permans
  - periodic paralysis
- Proximal myotonic myopathy
  - with hypothyroidism
  - others
- Chloride channel (uncommon)
- Rarely associated with pain
  - myotonic dystrophy
  - calcium channel periodic paralysis

Differential diagnosis – Disorders resembling myotonia

Myotonia, the delay in relaxation of muscle that results from repetitive depolarization of the sarcolemma, cannot be distinguished clinically from some forms of muscle contracture (Table 4). Contracture is, by definition, electrically silent. Typically intensely painful, muscle contractures are most commonly seen with substrate utilization defects: glycogenoses such as myophosphorylase deficiency and lipid storage disorders such as carnitine palmitoyl transferase (CPT) deficiency. Contractures are also characteristic of Brody's disease, and can be sufficiently painful to be the major symptom. The contractures of Brody's disease can be clinically indistinguishable from myotonia. Brody's disease can be suspected in practice by the finding of clinical "myotonia" that is not confirmed by EMG evidence for myotonia (see Karpati, this volume).

Table 4. Disorders of muscle relaxation

Disorder	Pathophysiology	Site of origin	Pain
Cramps	Repetitive motor unit-discharges	Motor neuron	Invariably intense
Tetany	Repetitive firing of nerve	Peripheral nerve	Occasionally severe
Neuromyotonia	Repetitive firing of nerve	Peripheral nerve	Typically slight
Myotonia	Repetitive firing of muscle fibers	Muscle	Occasionally severe
Contracture	Electrically silent failure of muscle to relax	Contractile apparatus	Often severe (occasionally painless)

Muscle cramps, characterized by intense focal pain, result from repetitive motor unit discharges [22]. Cramps are characteristic of denervating illnesses such as amyotrophic lateral sclerosis. Motor neuron diseases are more commonly a cause of cramps than peripheral neuropathy [22]. Tetany, seen with hypocalcemia or occasionally with hypomagnesemia can also occur with normal calcium and magnesium levels. Such cases of normocalcemic tetany are often called spasmophilia. Pain is often present with tetany but may be of only slight extent [22].

The pathophysiology of pain in myotonic disorders

Pain in the myotonic disorders is usually triggered or worsened by factors superimposed on the underlying genetic lesion [12, 15, 23] (Table 5). While the basis for pain in myotonic disorders is not fully elucidated, there are considerable data addressing the point. Although intense muscle contraction reduces muscle blood flow, there is little or no evidence that ischemia with or without muscle necrosis occurs. Where pain is triggered by an inciting factor such as potassium in sodium channel myotonia, it seems clear that myotonia itself causes the pain [14, 24].

Table 5. Factors precipitating pain in myotonic disorders

- Sodium channel myotonia
  - Potassium
  - Fasting
  - Exercise
  - Cold
- PROMM
  - Hypothyroidism
  - Exercise
  - Heat
- Chloride Channel Myotonia
  - Exercise

In some instances where only certain family members experience pain in association with myotonia, the pain may reflect coincidental disease in individuals reporting pain. Such a situation would be analogous to the symptoms in patients with "double trouble" (see Di Mauro, this volume). A second metabolic defect added to an underlying metabolic myopathy may produce unusually severe symptoms (Di Mauro, this volume).

Treating the symptoms of myotonic disorders and periodic paralyses

The symptoms of patients with the myotonias and periodic paralysis (Table 6) are now subject to molecular characterization and study. The molecular defect can be expressed and studied *in vitro* [3, 7-9]. Moreover, the ability to define the specific disorder permits natural history studies and the development of the treatment trials in a defined population [25].

There are no randomized controlled trials of treatment of most myotonic disorders and consequently no standardized treatment regimens. There was, however, a treatment trial of quinine and procainamide in the myotonic disorders - myotonia congenita and myotonic dystrophy [26]. This paper and subsequent anecdotal experience has suggested the treatments noted in Table 7.

The following illustrative cases highlight the occurrence of pain in myotonic disorders as well as emphasizing the questions surrounding the pathogenesis of pain.

Table 6. Major symptoms and signs in Myotonic diseases and periodic paralysis

Episodic weakness
Fixed (non-episodic) weakness
Myotonia - stiffness
Myotonia - pain
Cardiac arrhythmias

Table 7. Treatment strategies for Muscle Pain in myotonia

• Sodium channel
- acetazolamide
- dichlorphenamide
- mexiletine
• PROMM – uncertain
• Chloride channel
- mexiletine (recessive)
- quinine (dominant)
• Myotonic dystrophy
- phenytoin

Case 1

A 36-year-old woman complained of "back pain" for three years. In retrospect, the pain may have been present for 10-15 years previously. She was a school teacher who continued working full time despite the pain. The pain occurred only with activity such as standing or walking for 15-30 minutes and would abate with rest. It was not progressive and never occurred at rest. Physical examination was normal. Thorough MRI and radiographic evaluation of the spine was normal.

After numerous evaluations by orthopedists and neurologists, the patient was seen for an additional neurologic opinion and had repeat EMG. Although previously normal (by report), EMGs of proximal muscle demonstrated widespread, though variable, myotonia in numerous muscles.

Re-examination of muscle strength disclosed slight (grade 4 MRC) weakness of neck flexion and of biceps muscles. Ophthalmologic slit lamp exam identified the characteristic cataracts of myotonic dystrophy but trinucleotide repeat testing for the gene lesion of myotonic dystrophy was normal. Assessment of family indicated myotonia (by EMG) and cataracts in other family members in an autosomal dominant pattern. A diagnosis of PROMM was considered likely.

Comment. This patient illustrates an important issue with regard to the diagnosis of PROMM. Until the gene(s) responsible for PROMM is cloned, it is unclear whether or not PROMM is a single disease entity. If one defines PROMM as any autosomal dominant disorder with myotonia [1] and weakness [2], without the systemic signs typical of PROMM, it seems possible that one will include patients with myotonia caused by different gene lesions. The discovery of gene lesion or linkage is necessary to address this issue. The presence of cataracts in a young person makes PROMM more defensible but does not necessarily argue for a single disease.

Case 2

A 35-year-old man had a history of thyroid cancer at age 13 and was maintained on daily L-thyroxine 0.3-0.25mg. Inadvertently (by pharmacy error) he received a ten-fold lower dose. After months of this erroneous treatment he developed symptoms of muscle pain, cramping, chest pain and fatigue. On assessment, T4 was undetectable, TSH was > 50 IU/ml and CK values were increased by 60-fold (>10,000 IU/L). His dosage of L-thyroxine was returned to his previous dosage and some of his symptoms improved. However, despite long-term restoration of the euthyroid state he complained of cramps in his legs, muscle aches and fatigue. He also had difficulty with fine motor control and releasing objects. Because of these persistent symptoms, he has underwent a thorough neurologic evaluation. Clinical and percussion myotonia and slight (grade 4 MRC) neck flexor, deltoid and hip muscle weakness were present. EMG revealed myotonic discharges in the biceps, deltoid, thumb adductor, quadriceps and tibialis anterior muscles. Muscle biopsy of the quadriceps femoris muscle showed increased variability in fiber size, type 1 fiber atrophy, type II fiber hypertrophy, and increased numbers of central nuclei (in 40% of fibers) and pyknotic nuclear clumps. His CK remained two-fold increased. Genetic analysis shows a normal size of CTG repeats for the DM gene. Past medical history is significant for oligospermia. His 65-year-old father (II-4) and 74-year-old paternal uncle (II-5) have complained of difficulty climbing a flight of stairs, getting up from a squat, stiffness and difficulty releasing objects since the age of 40.

Comment. Hypothyroidism is a relatively common medical problem and it is conceivable that the association between the development of worsening signs of PROMM and hypothyroidism that this patient had is coincidental. However, we have seen two additional patients whose symptoms of muscle disease only began when they developed hypothyroidism. Prior to the diagnosis of hypothyroid myopathy all three patients were asymptomatic and had no functional limitations. It seems likely that all three patients had subclinical PROMM that emerged abruptly during the onset of severe hypothyroidism.

The future

Studies of large numbers of patients with genotype/phenotype correlations may determine if specific mutations are consistently characterized by pain. *In vitro* expression of specific mutations may permit quantitation of the electrophysiologic defect produced by a given mutation. The effect of environmental milieu can also be assessed in *in vitro* models. Quantitative protocols for the study of muscle pain and myotonia will facilitate controlled trials of treatment with each putative therapy [25].

References

1. Brown KA, al-Gazali LI, Moynihan LM et al (1997) Genetic heterogeneity in Schwartz-Jampel syndrome: two families with neonatal Schwartz-Jampel syndrome do not may to human chromosome 1p34-p36.1. J Med Genet 34 (8):685-687
2. Cannon SC (1997) From mutation to myotonia in sodium channel disorders. Neuromuscular Disorders 7 (4):241-249
3. Cannon SC (1996) Ion-channel defects and aberrant excitability in myotonia and periodic paralysis. Trends in Neurosciences 19 (1):3-10
4. Griggs RC, Tawil R, Brown RJ et al and the Working Group on Periodic Paralysis (1996) Implications of molecular defects for classification and treatment of periodic paralysis. In: Kimura J, Shibasaki H (eds) Xth International Congress of EMG and Clinical Neurophysiology. Kyoto, Elsevier Science BV pp 160-169
5. Ptáček LJ (1997) Channelopathies: ion channel disorders of muscle as a paradigm for paroxysmal disorders of the nervous system. Neuromuscular Disorders 7:250-255
6. Ptáček LJ, Johnson KJ, Griggs RC (1993) Mechanisms of disease: Genetics and physiology of the myotonic muscle disorders. New Eng J Med 328:482-489
7. De Luca A, Pierno S, Natuzzi F et al (1997) Evaluation of the antimyotonic activity of mexiletine and some new analogs on sodium currents of single muscle fibers and on the abnormal excitability of the myotonic ADR mouse. J Pharmacol & Experimental Therapeutics 282 (1):93-100
8. Elbaz A, Vale-Santos J, Jurkat-Rott K et al (1995) Hypokalemic periodic paralysis and the dihydropyridine receptor (CACNL1A3): genotype phenotype correlations for two predominant mutations and evidence for the absence of a founder effect in 16 caucasian families. Am J Hum Genet 56 (2):374-380
9. Fahlke C, Beck CL, George AL Jr (1997) A mutation in autosomal dominant myotonia congenita affects pore properties of the muscle chloride channel. Proceedings of the National Academy of Sicences of the United States of America 94 (6):2729-2734
10. Ptáček LJ, Griggs RC (1996) Familial Periodic Paralysis. In: Andreoli TE, Brown A, Fambrough D, Hoffman J, Schultz S, Welsh M (eds) The Molecular Biology of Membrane Transport Disorders. Plenum Publishing, New York, pp 625-642
11. Tawil RT, Ptáček LJ, Pavlakis SG et al (1994) Andersen's syndrome: Potassium-sensitive periodic paralysis, ventricular ectopy and dysmorphic features. Ann Neurol 35:326-330
12. Lehmann-Horn F, Mailander V, Heine R et al (1995) Myotonia levior is a chloride channel disorder. Human Molec Genetics 4 (8):1397-1402
13. Sunohara N, Tomi H, Nakamura A et al (1996) Myotonia congenita with painful muscle cramps. Internal Medicine 35 (6) 507-511
14. Ptáček LJ, Tawil R, Griggs RC et al (1992) Linkage of atypical myotonia congenita to a sodium channel locus. Neurology 42:431-433

15. Ptáček LJ, Tawil R, Griggs RC et al (1994) Sodium channel mutations in acetazolamide-responsive myotonia congenita, paramyotonia congenita and hyperkalemic periodic paralysis. Neurology 44:1500-1503
16. Thornton CA, Griggs RC, Moxley RT III (1994) Myotonic dystrophy with no trinucleotide repeat expansion. Ann Neurol 35:269-272
17. Udd B, Krahe R, Wallgren-Pettersson C et al (1997) Proximal myotonic dystrophy – a family with autosomal dominant muscular dystrophy, cataracts, hearing loss and hypogonadism: heterogeneity of proximal myotonic syndromes? Neuromuscular Disorders 7(4):217-228
18. Fontaine B, Nicole S, Topaloglu H et al (1996) Recessive Schwartz-Jampel syndrome (SJS): confirmation of linkage to chromosome 1p, evidence of genetic homogeneity and reduction of the SJS locus to a 3 -cM interval. Human Genetics 98 (3):380-385
19. Ptáček LJ, Tawil R, Griggs RC et al (1994) Dihydropyridine receptor mutations cause hypokalemic periodic paralysis. Cell 77:863-868
20. Rose M, Griggs RC, Tawil R (1998) Channelopathies. In. Pulst SM (ed) Neurogenetics. Contemporary Neurology Series, FA Davis Company (in press)
21. Sansone V, Griggs RC, Meola G et al (1997) Andersen's syndrome: A distinct periodic paralysis. Ann Neurol 42:305-312
22. Griggs RC (1998) Episodic muscle spasms, cramps, and episodic weakness. In: Fauci AS, Braunwald E, Isselbacher KJ et al (eds) Harrison's Principles of Internal Medicine, The McGraw-Hill Companies, 14th edn, pp 118-122
23. Sander HW, Tavoulareas GP, Chokroverty S (1996) Heat-sensitive myotonia in proximal myotonic myopathy. Neurology 47:956-962
24. Ptáček LJ, George AL, Barchi RL et al (1992) Mutations in an S4 segment of the adult skeletal muscle sodium channel cause paramyotonia congenita. Neuron 8:891-897
25. Griggs RC (1994) Developing new treatments for muscle disease: prospects and promise. Current Opinion in Neurology 7:422-426
26. Munsat TL (1967) Therapy of myotonia. A double-blind evaluation of diphenylhydantoin, procainamide, and placebo. Neurology 17(4):359-367
27. Wollnik B, Kubisch C, Steinmeyer K et al (1997) Identification of functionally important regions of the muscular chloride channel CIC-1 by analysis of recessive and dominant myotonic mutations. Human Molecular Genetics 6(5):805-811
28. Zhang J, George, AL Jr, Griggs RC et al (1996) Mutation in the human skeletal muscle chloride channel gene (CLCN1) associated with dominant and recessive myotonia congenita. Neurology 47:993-998

Exertional Compartmental Syndromes

R.G. Haller

Compartmental syndrome denotes a condition in which an increase in the volume of tissues in an anatomically closed space results in an abnormal rise in tissue and intracompartmental pressure that compromises the circulation and the function of tissues in that space [1]. Exertional compartmental syndrome indicates that the mechanism of increased intracompartmental pressure involves exercise. Acute and chronic exertional compartmental syndromes have been described. In acute syndromes the pathological rise in intracompartmental pressure that is the central mechanism of the syndrome is attributable primarily to muscle injury with subsequent tissue swelling occurring as a consequence of muscle necrosis and edema. Patients with inborn errors of muscle energy metabolism who are subject to recurrent exertional rhabdomyolysis are liable to recurrent acute compartment syndrome. In individuals with chronic exertional compartment syndrome, an abnormal rise in intracompartmental pressure accompanies physiological fluid shifts that are induced by metabolic and hydrostatic effects of exercise [2]. The exaggerated rise in compartment pressure presumably results from a "tight" compartment in relation to the degree of tissue expansion.

The *sine qua non* of exertional compartmental syndromes is an abnormal increase in intracompartmental pressure that occurs as a result of musculoskeletal pathology (muscle necrosis) or physiologic fluid shifts in an insufficiently compliant anatomic space; and the 'gold standard' for diagnosis of acute and especially chronic compartment syndromes is direct measurement of compartmental pressure [3, 4] The precise mechanisms by which increased compartment pressure mediates cellular dysfunction and injury is incompletely understood, but the leading hypothesis is that increased tissue pressure within the affected compartment impairs nutrient blood flow by reducing the arterio-venous pressure gradient across tissues contained within the compartment [5]. Local blood flow (LBF) is directly related to the arterio-venous pressure gradient (PA-PV) and is inversely proportional local vascular resistance (R). Increased tissue pressure compresses low pressure venules, effectively increasing pressure in intracompartmental veins. As a result the arterio-venous pressure gradient is decreased. Experimental studies indicate that capillary blood flow ceases when external pressures reach a level of 25-30 mmHg below mean arterial pressure [6]. Ischemia compromises muscle metabolism, and if sufficiently severe or prolonged ultimately results in muscle necrosis. Edema accompanying muscle necrosis in turn promotes a further rise in intracompartmental pressure. Measurement of compartmental blood flow generally supports this pathophysiological scheme, demonstrating impaired blood flow in affected individuals and improved blood flow after surgical treatment [7-9].

Anatomic considerations

Virtually any muscle that is invested by relatively non compliant fascial and/or osseous tissue may be involved in an exertional compartment syndrome, but compartments of the lower leg are most commonly affected. The lower leg contains four anatomic compartments: anterior, lateral, and deep and superficial posterior compartments (Fig. 1). The anterior compartment most commonly is involved, probably related to a high degree of inelasticity of this space. The anterior compartment is bounded by the tibia medially, the interosseous membrane (between the tibia and fibula) posteriorly, the anterior crural septum laterally, and superficial fascia anteriorly. Contained within this compartment are the tibialis anterior, extensor hallicus longus, extensor digitorum longus, and peroneus tertius muscles, the deep peroneal nerve and anterior tibial vessels. Symptoms of acute anterior

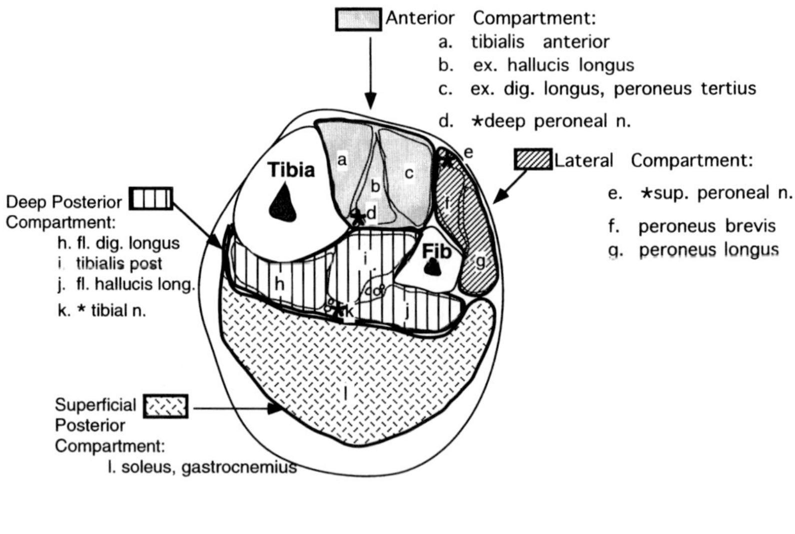

Fig. 1. Anatomic compartments of the lower leg

compartment syndrome include pain and swelling diffusely over the antero-lateral leg with pain often radiating to the ankle. Compression or ischemia of the deep peroneal nerve with advancing anterior compartment syndrome causes weakness of the foot extensors and paresthesiae or loss of sensation on the dorsum of the foot between the first and second toes. Progression may result in paralysis of muscles in the anterior compartment and ultimately fixed contractures involving these muscles.

Rarely involved alone, but often in combination with the anterior compartment is the lateral compartment of the lower leg. The lateral compartment is bounded by the fibula and the anterior crural septum medially, by the posterior crural septum posteriorly, and by superficial fascia anterolaterally. The lateral compartment contains the peroneus brevis and peroneus longus muscles and the superficial peroneal nerve. Symptoms are similar to those of anterior compartment syndrome. Involvement of the superficial peroneal nerve is indicated by weakness of foot eversion and by sensory loss over the entire dorsum of the foot.

The deep posterior compartment is bounded by the interosseous membrane anteriorly, the tibia medially, the fibula and posterior crural septum laterally, and by the intermuscular septum posteriorly. This compartment contains the tibialis posteriori flexor digitorum longus, and flexor hallicus longus muscles, the posterior tibial and peroneal vessels, and the tibial nerve. Pain emanating from the posterior compartment typically occurs along the posteromedial border of the tibia. Involvement of the tibial nerve in the deep posterior compartment may result in sensory loss in the sole of the foot and weakness of foot flexion and adduction. The superficial posterior compartment containing the soleus, gastrocnermius and plantaris muscles is rarely involved in compartment syndrome.

Acute exertional compartment syndrome

Exercise-related compartmental syndrome initially was described in otherwise healthy military recruits who, after prolonged marching, developed acute anterior compartment syndrome with pain and swelling of the anterior leg that progressed to paralysis of foot dorsiflexion compatible with peroneal palsy. The pathologic basis of these symptoms was demonstrated to be ischemic necrosis of the muscles of the anterior compartment, i.e. the tibialis anterior and long extensors of the toes [10, 11]. A direct relationship between marching and the ischemic muscle injury was recognized in the term that was applied to this syndrome - "march gangrene" - but the pathogenic role of exercise was incompletely understood. It was postulated that exercise promoted tissue injury by causing metabolic fatigue and by increasing circulatory demand when blood flow was compromised. Current understanding suggests that the triggering role of exercise likely involves mechanical injury of muscle.

Otherwise healthy subjects are liable to exertional muscle injury with or without associated compartmental syndrome under circumstances of unaccustomed, severe exercise. Exercise involving repetitive forceful *eccentric* muscle contractions typically is involved. Such exercise involves lengthening muscle contractions under circumstances in which affected muscles act as 'shock absorbers' such as down-hill running. By forcefully stretching contracted muscle, such exercise produces mechanical injury to muscle cells resulting in tears in the sarcolemma and in disruption of contractile elements that may progress to muscle necrosis [12-14] The clinical features of such injury usually evolve over a longer time frame than other forms of rhabdomyolysis with pain and swelling that peak 24-48 hours after the injury-producing exercise. A similar mechanism of muscle injury presumably underlies the common phenomenon of muscle soreness evolving 1-2 days after unaccustomed exercise that has been termed "delayed onset muscle soreness." Military recruits engaged in rigorous physical training and long sedentary individual imbued with a zeal to achieve fitness through 'no pain no gain' exercise are subject to severe cases of this form of muscle injury and may develop frank rhabdomyolysis with myoglobinuria [15, 16]. Symptoms initially may be minor, but evolve progressively 12-48 hours later. When pigmenturia is associated, its onset commonly is similarly delayed 24-48 hours after the initial injury. Hughes [11] emphasized a similar delay in the evolution of symptoms in acute anterior compartment syndrome consistent with this mechanism of muscle injury. The susceptibility of the muscles of the anterior compartment of the leg in march gangrene may relate to the fact that these muscles are subject to repetitive eccentric contractions during heel strike, especially during forced marching or running. Muscle swelling produced by such injury initiates the vicious cycle of increased intracompartmental pressure, impaired blood flow, and ischemic necrosis that produces the full blown syndrome.

Acute compartment syndrome threatens the viability of muscle and neurovascular components of the affected tissue space and risks permanent loss of function of its contents as was the outcome in initially described cases [10, 11]. Thus, accurate diagnosis and timely surgical intervention when required to prevent permanent injury are essential. The clinical diagnosis is based upon a typical history of symptoms, occurring hours to days after severe exercise, consisting of unremitting pain and progressive increases in compartment tightness that limits the mobility of the affected limb. Paresthesiae or loss of sensation indicates pressure or ischemia involving neural elements in the affected compartment. Physical findings consistent with the diagnosis include: 1) tenseness within the fascial boundaries of the compartment; 2) pain that seems out of proportion to the clinical situation; 3) extreme pain on attempted passive movement of the muscles within the compartment; 4) hypesthesia involving nerves that traverse the compartment, and 5) weakness of muscles of the compartment. Variables which affect the outcome of acute compartment syndromes include the level of increase in compartmental pressure as well as the duration of pressure. The decision to employ surgical decompression often is made on the basis of the evolution of clinical symptoms and signs, particularly the emergence of neural involvement. Direct measurement of intracompartmental pressure may be employed to determine the need for surgical intervention. In a series including 25 patients with acute anterior compartment syndrome, Matsen and coworkers [17] found that individuals with maximum intracompartmental pressures of 45 mmHg or less did not require fasciotomy and demonstrated no permanent sequelae on follow up. In contrast all patients with maximum compartment pressures of 55 mmHg or greater experienced significant neuromuscular deficits. However, patients vary in their tolerance of compartmental pressure and surgeons employ differing guidelines. For example, Rorabeck [18] and Mubarak [19] have recommended decompression for resting pressures in the range of 30-40 mmHg.

Recurrent acute compartment syndrome

Patients subject to recurrent exertional rhabdomyolysis attributable to inborn errors of carbohydrate, lipid, or mitochondrial metabolism may experience recurrent rhabdomyolysis and may be subject to recurring compartment syndrome (Table 1). Individuals with muscle glycolytic defects are particularly prone to this complication. Such patients develop muscle injury with brief exertion in activities of daily living and may experience severe focal muscle injury with brief vigorous exertion [20]. Swelling of injured muscle within a confined anatomic space may transform a relatively bland injury into a frank compartment syndrome. Patients with muscle phosphorylase deficiency (McArdle's disease), for example, may experience severe rhabdomyolysis after activities such as brief sprinting or holding a heavy object and, as a result, may develop elevations of serum CK of 200 fold the upper limit of normal. Despite laboratory evidence of severe muscle injury, such patients often experience relatively minor symptoms of muscle aching. At other times, and particularly with focal injury affecting the muscles of the shoulder girdle or back, exertional muscle injury may produce excruciating pain.

Typically under these circumstances, there is a delay of some hours between the injury and the onset of severe, unremitting pain consistent with the development of necrosis-induced local muscle swelling within a restricted anatomic space. One of my patients has undergone fasciotomy after developing an acute exertional anterior compartmental syndrome related to brief sprinting.

Table 1. Hereditary disorders associated with recurrent exertional rhabdomyolysis

Defects (or Disorders)	Type of exercise
• **Carbohydrate defects**	
- myophosphorylase	
- phosphofructokinase	
- phosphoglycerate mutase	brief, intense - sprinting, weight-lifting
- phosphoglycerate kinase	
- lactate dehydrogenase	
• **Lipid defects**	
- carnitine palmitoyl transferase II	
- very long chain acyl CoA dehydrogenase	prolonged, submaximal - especially when combined with fasting
- trifunctional protein	
- short-chain hydroxyacyl CoA dehydrogenase	
• **Mitochondrial defects**	
- multiple iron-sulfur proteins (SDH/aconitase)	
- coenzyme Q deficiency	intense exercise,? other factors
- mtDNA – large scale rearrangements, point mutations (rare)	

Chronic exertional compartment syndrome

Chronic exertional compartment syndrome is a disorder encountered most commonly by orthopedic surgeons and specialists in sports medicine, and only rarely by neurologists. Athletes, particularly runners, are most commonly affected. Affected individuals are free of symptoms at rest, but exercise, after a variable interval, results in symptoms of pain, tightness, and sometimes paresthesiae that ultimately limits exercise.

Cessation of exercise results in gradual resolution of symptoms. Symptoms usually arise from the anterior compartment of the leg, but the lateral, and deep posterior compartments also may be involved [21-23].

The fundamental mechanism of symptoms involves an exaggerated increase in tissue pressure in the affected compartment that ultimately limits muscle blood flow. Compartment pressures often are normal or minimally elevated at rest. But with exercise, there is an exaggerated rise in compartment pressure and after exercise the decline in compartment pressure is delayed (Fig. 2) [4]. These abnormal changes in compartment pressure with exercise presumably relate to: 1) fluid shifts from the vascular to the muscle and interstitial spaces, and 2) an abnormally incompliant compartment.

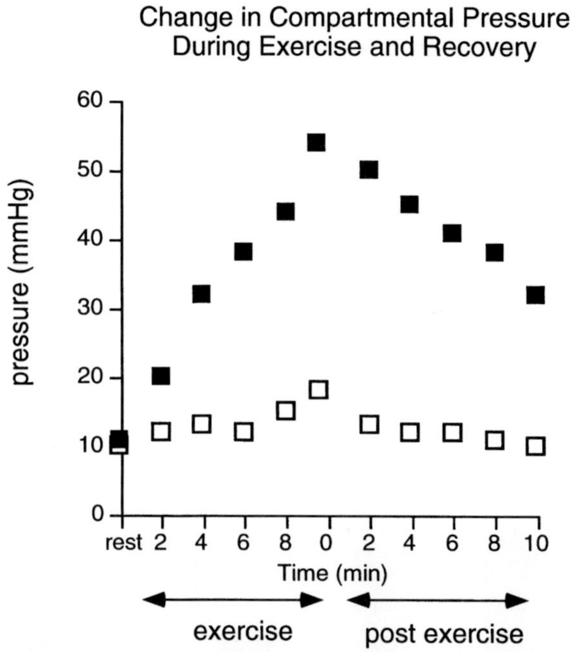

Fig. 2. Representative measurements of pressures in the anterior compartment of the leg in a patient with chronic exertional anterior compartment syndrome (*closed squares*) and in an unaffected individual (*open squares*). Note that resting compartment pressures are similar, but with exercise the affected individual has an exaggerated increase in compartment pressures. Also, at the end of exercise, the decline in compartment pressures is delayed in the patient with chronic compartment syndrome

Heavy exercise normally produces a shift in water from the vascular compartment to muscle intracellular and interstitial spaces. This phenomenon is marked by increasing concentrations of cellular and protein components of blood as well as increased muscle content of water consistent with the movement of free water from vascular to extravascular spaces. Correspondingly, muscle volume increases, accounting for the swelling ("pumping up") of muscle after heavy exercise. Sjøgaard and Saltin [2] found that brief heavy bicycle exercise resulted in an approximately 15% increase in muscle water. Correspondingly there was a decrease in intravascular water as indicated by a 19% increase in the concentration of plasma protein and a 14% increase in hematocrit [2]. The mean total water content of vastus lateralis, a muscle heavily recruited in cycle exercise, increased from 313 ml/100 g dry wt at rest to 359 ml/100 g dry wt after exercise (Fig. 3) whereas the water content of muscle that was inactive in cycle exercise (triceps) was unchanged. Using tritiated inulin to distinguish intracellular and extracellular muscle water compartments, these investigators determined that almost 60% of the increase in vastus lateralis muscle water was attributable to an increase in extracellular (interstitial) muscle water which rose from 33 ml/100 g dry wt (i.e. 10.5% of total muscle water) at rest to 60 ml/100 g dry wt (16.7% of total muscle water) after maximal cycle exercise. Thus, extracellular muscle water increased more than 80% from rest to exercise whereas the calculated increase in intracellular muscle water was only 6.8% (Fig. 3). This increase in muscle, primarily extracellular, water is detectable as an increase in proton magnetic resonance image intensity attributable to a prolongation of proton T2 relaxation time [24, 25].

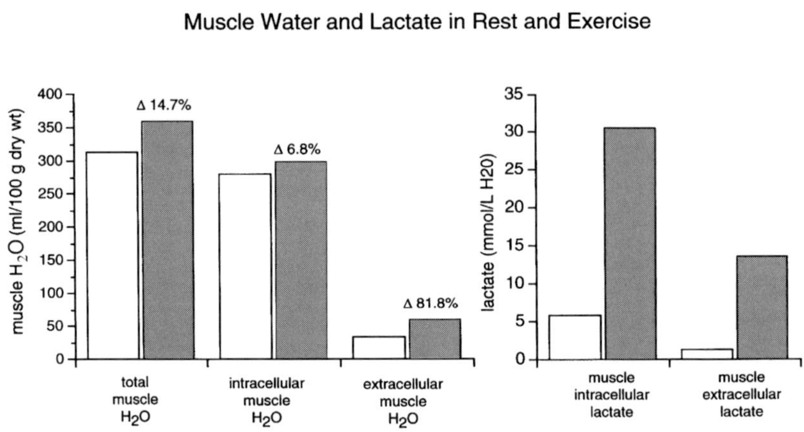

Fig. 3. Muscle intracellular and extracellular (interstitial) water (right portion of figure) and lactate (left) at rest (open bars) and after (shaded bars) intense bicycle exercise. Data adapted from [2]

The physiologic basis of this increase in muscle water involves osmotic and hydrostatic changes that accompany exercise. The major change in tissue osmolarity in heavy exercise is attributable to an increase in muscle lactate. Sjøgaard and Saltin [2] found that muscle intracellular lactate rose from 5.7 mM at rest to 30.6 mmol/L after heavy exercise while muscle interstitial lactate levels increased from 1.3 to 14 mmol/L (Fig. 2). Based on the higher intracellular compared to extracellular lactate, the osmotic gradient would seemingly favor a greater accumulation of intracellular water. The fact that the opposite was the case, i.e. that the increase in extracellular water was greater than the increase in intracellular water, suggested that additional, presumably hydrostatic variables contributed to the preferential extracellular water accumulation [2]. However, Fleckenstein and coworkers [26] found that muscle water accumulation after exercise as monitored by MRI does not occur when muscle lactate production is blocked due to muscle phosphorylase deficiency, implying that osmotic factors are the primary determinant of such fluid shifts. Based on the results of xenon clearance, Jehenson and coworkers [27] have suggested that the exercise increase in muscle blood flow may be impaired in McArdle's disease and account for the impaired increase in muscle water with exercise. However, this interpretation is contradicted by plethysmographic and thermal dilution measurements which have found muscle blood flow to be exaggerated in McArdle patients compared to healthy subjects [28-30].

There is some evidence that the level of increase in muscle water may be exaggerated in patients with chronic exertional compartment syndrome as a consequence of recurrent mechanical injury of muscle or to greater than normal increases in muscle lactate levels [7]. In addition, several investigators have noted that affected individuals have evidence of an abnormally tight compartment as indicated by an incidence of muscle hernias ranging from 40-60% [4, 7, 22, 31]. In such patients surgical correction of the hernia is contraindicated and may, in fact, precipitate an acute compartment syndrome [11]. Muscle hypertrophy also has been cited as a possible factor contributing to chronic compartment syndrome in athletes [7, 21].

The diagnosis of chronic exertional compartment syndrome is based upon a compatible history and evidence of increased compartmental pressure with exercise that normally provokes symptoms and by the finding of a prolonged return to baseline levels. Pedowitz and coworkers [4] reviewed the records of 120 patients referred with symptoms suggesting possible chronic compartment syndrome of the leg. Forty-five (37%) were diagnosed as having a chronic compartment syndrome, whereas 75 (63%) had the syndrome excluded by compartment pressure measurements before and after exercise. The only significant difference on history and physical examination was a higher incidence of muscle herniae in the chronic compartment syndrome patients (46%) compared to those without the syndrome (13%). Intramuscular pressures considered diagnostic of chronic compartment syndrome were: 1) pre-exercise pressure $\geq$ 15 mmHg; 2) a 1 min post exercise pressure of $\geq$ 30 mmHg, or 3) a 5 min post exercise pressure of $\geq$ 20 mmHg. In both groups of patients, symptoms were most commonly referred to the anterior compartment, were typically unilateral (though 15-20% of cases complained of bilateral symptoms), and were usually brought on by running.

In patients with typical symptoms and elevated exercise pressures, fasciotomy commonly relieves symptoms [32]. In the series of Pedowitz et al. [4], 60% of the patients with chronic compartment syndrome were able to resume exercise at the highest level that was possible before the onset of symptoms. However, one third of patients continued to have symptoms of moderate to severe exertional pain. Schepsis and coworkers [33] reported a much higher incidence of "excellent" results in patients with chronic anterior compartment syndrome (96%) than in those with chronic deep posterior compartment syndrome (25%). One of the most convincing studies of the pathophysiology of chronic exertional compartment syndrome and of the benefits fasciotomy was reported by Qvarfordt and coworkers [7]. Of 108 patients evaluated for lower leg pain with resting and post exercise pressure recordings of the anterior compartment, these investigators identified 15 subjects (14%) with increased pressures compatible with chronic anterior compartment syndrome and 93 with normal pressures, in whom the syndrome was excluded. Using the wick catheter technique to continuously monitor compartment pressure at rest, during and after exercise, the 15 patients were found to have: 1) a mean resting compartment pressure of 22 mmHg (range 18-28) compared to a normal mean of 8 mmHg (range 2-23); 2) exercise pressures that rose to a mean of 80 mmHg (range 48-135) compared to a normal mean of 26 (range 8-60), and 3) a prolonged time required for post exercise compartment pressure to decrease to half way to the initial resting pressure (T1/2) of 6±3 minutes compared to a normal value of 3 seconds. Fourteen of 15 patients were symptom free after fasciotomy. Correspondingly, resting, exercise, and post exercise pressure recordings determined 3 months after surgery were normal. Fasciotomy had no effect on resting muscle blood flow but exercise blood flow was 36% higher after surgery [7].

The measurement of compartmental pressures is invasive and cumbersome. Thus, it would be desirable to identify a reliable non-invasive means of differentiating exertional compartment syndrome from other mechanisms of exertional pain. Eskelin and coworkers [34] reported a direct correlation between compartment pressure measurements and magnetic resonance signal intensity in subjects with suspected anterior compartment syndrome and suggested that MR imaging after exercise could assist in diagnosis. However, the data presented suggest that MRI differences were subtle in some patients who had clear cut pressure elevations; and Fleckenstein [35] has cautioned that variables other than compartmental pressure changes can influence increased MR image intensity with exercise. Another tool that recently has been employed for non invasive evaluation of possible chronic exertional compartment syndrome is near infrared spectroscopy [9, 36]. The technique is based upon the capacity of light in the NIR spectrum (700-1000 nm) to penetrate tissues to a depth of several centimeters and upon the differential NIR absorption characteristics of oxygenated hemoglobin + myoglobin compared to reduced hemoglobin + myoglobin [37]. This permits the qualitative assessment of net extraction of O_2 relative to O_2 delivery by the circulation over the area of tissue sampled by the device at rest, during exercise, and with recovery from exercise. Any process that impairs the increase in delivery of oxygenated blood relative to its utilization in exercise, including an exaggerated increase in compartmental pressure, results in a net reduction in tissue oxygenation. Mohler and coworkers [9] found that patients with chronic anterior compartment syndrome had a greater than normal decrease in levels of tissue oxygenation and an abnormally delayed return to baseline post exercise consistent with impaired muscle blood flow during exercise and recovery. While further evaluation of these results are warranted, this result suggests a possible role of NIR in the diagnosis of this condition.

References

1. Matsen FAI (1980) Compartment syndromes. Grune & Stratton Inc New York
2. Sjøgaard G, Saltin B (1982) Extra- and intracellular water spaces in muscles of man at rest and with dynamic exercise. Am J Physiol 243:R271-R280
3. Mubarak SJ (1983) A practical approach to compartmental syndromes. II. Diagnosis. AAOS Instr Course Lect 32:92-102
4. Pedowitz RA, Hargens AR, Mubarak SJ, Gershuni DH (1990) Modified criteria for the objective diagnosis of chronic compartment syndrome of the leg. Am J Sports Med 18:35-40
5. Matsen FAI (1983) A practical approach to cornpartmental syndromes. 1. Definition, theory, and pathogenesis. AAOS Instr Course Lect 32:88-92
6. Feneman RS, Slaaf DW, Lindbom L, Tangelder GJ, Arfors KE (1981) Muscle bloodflow disturbances in compartment syndromes and the role of elevated total muscle-tissue pressure in these disturbances. In: Hargens AW (ed) Tissue fluid pressure and composition. Williams and Wilkins, Baltimore
7. Qvardordt P, Christenson JT, Eklöf B, Ohlin P, Saltin B (1983) Intramuscular pressure, muscle blood flow, and skeletal muscle metabolism in chronic anterior tibial compartment syndrome. Clin Orthopedics 179:284-290
8. Takebayashi S, Takazawa H, Sasaki R, Miki H, Soh R, Nishimura J (1997) Chronic exertional compartment syndrome in lower legs: localisation and follow-up with thallium-201 SPECT imaging. J Nucl Med 38:972-976
9. Mohler LR, Styf JR, Pedowitz RA, Hargens AR, Gershuni DH (1997) Intramuscular deoxygenation during exercise in patients who have chronic anterior compartment syndrome of the leg. J Bone Joint Surg 79A:844-849
10. Horne CE (1945) Acute ischaemia of the anterior tibial muscle and the long extensor muscles of the toes. J Bone Joint Surg 27:615-622
11. Hughes JR (1948) Ischaemic necrosis of the anterior tibial muscles due to fatigue. J Bone Joint Surg 30B:581-594
12. Armstrong RG, Warren GL, Warren JA (1991) Mechanisms of exercise-induced muscle fibre injury. Sports Medicine 12:184-207
13. McNeil PL, Khakee R (1992) Disruption of muscle fiber plasma membranes. Role in exercise-induced damage. Am J Pathol 140:1097-1109
14. McNeil PL, Steinhardt RA (1997) Loss, restoration, and maintenance of plasma membrane integrity. J Clin Invest 137:1-4
15. Demos MA, Gitin EL, Kagen LJ (1974) Exercise myoglobinemia and acute exertional rhabdomyolysis. Arch Intern Med 134:669-673
16. Knochel JP (1993) Mechanisms of rhabdomyolysis. Curr Op Rheumatol 5:725-73 1
17. Matsen FAI, Winquist RA, Krugmire RB (1980) Diagnosis and management of compartmental syndromes. J Bone Joint Surg 62-A: 286-291
18. Rorabeck CH, Castle GSP, Hardie R (1981) Compartment pressure measurements: an experimental investigation using the slit catheter. J Trauma 21:446
19. Mubarak SJ, Hargens AR (1981) Compartment syndromes and Volkmann's contracture. W.B. Sanders Co, Philadelphia

20. Fleckenstein JL, Peshock RM, Lewis SF, Haller RG (1989) Focal muscle injury and atrophy in human muscle glycolytic disorders. Muscle & Nerve 12:849-855
21. Styf J (1989) Chronic exercise-induced pain in the anterior aspect of the lower leg: an overview of diagnosis. Sports Medicine 7:331-339
22. Martens MA, Moeyersoons JP (1990) Acute and recurrent effort-related compartment syndrome in sports. Sports Medicine 9:62-68
23. Hutchinson MR, Ireland ML (1994) Common compartment syndromes in athletes – treatment and rehabilitation. Sports Med 17:200-208
24. Fleckenstein JL, Parkey RW, Peshock RM (1988) Acute effects of exercise on MRI of skeletal muscle in normal volunteers. AJR 151:231-237
25. Fleckenstein JL, Crues JVI, Reimers CD (1996) Muscle imaging in health and disease. Springer, NewYork, p 461
26. Fleckenstein JL, Haller RG, Lewis SF, Parkey RW, Peshock RM (1991) Absence of exercise induced MRI enhancement of skeletal muscle in MeArdle's disease. J Appl Physiol 71:961-969
27. Jehenson P, Leroy-Willig A, de Kerviler E, Merlt P, Duboc D, Syrota A (1995) Impairment of the exercise-induced increase in muscle perfusion in McArdle's disease. Eur Nuclear Med 22:1256-1260
28. McArdle B. Myopathy due to a defect in muscle glycogen breakdown. Clin Sci 10:13-33
29. Barcroft H, Greenwood B, McArdle B, McSwiney R, Semple SJG, Whelan RF, Youlten LJF (1966) The effect of exercise on forearm blood flow and on venous blood pH, pCO2 and lactate in a subject with phosphorylase deficiency in skeletal muscle. J Physiol 189: 44P-46P
30. Porte D Jr, Crawford DW, Jennings JB, Aber C, Mcllroy MB (1966) Cardiovascular and metabolic responses to exercise in a patient with McArdle's syndrome. N Eng J Med 275:406-412
31. Reneman RS (1968) The anterior and the lateral compartment syndrome of the leg. The Hague, Mouton
32. Rorabeck CH (1983) A practical approach to compartmental syndromes. III. Management. AAOA Instr Course Lect 38:102-113
33. Schepsis AA, Martini D, Corbett M (1993) Surgical management of exertional compartment syndrome of the lower leg; long-term follow up. Am J Sports Med 21:811-817
34. Eskelin MKK, Lötjönen JNP, Mäntysaari MJ (1998) Chronic exertional compartment syndrome: MR Imaging at 0.1 T compared with tissue pressure measurement. Radiology 206:333-337
35. Fleckenstein JL (1998) (Dys-)Functional MR imaging of skeletal muscle: a cautionary note. Radiology 206:305-307
36. Breit GA, Gross JH, Watenpaugh DE, Chance B, Hargens AR (1997) Near-infrared spectroscopy for monitoring of tissue oxygenation of exercising skeletal muscle in a chronic compartment syndrome model. J Bone Joint Surg 79A:838-843
37. Haller RG (1994) Oxygen utilization and delivery in metabolic myopathies. Ann Neurol 36:811-813

Reconciling Language Differences in Describing States of Muscle Shortening: Muscle Contracture and Related Conditions

G. Serratrice, L.P. Rowland

In his chapter on the classification of contracture, Rowland [1] started with the dictionary definition of contracture as "resistance to passive movement", without comment about muscle action potentials. He noted, however, that the word is also used to describe shortening of muscle without the generation of a muscle action potential, as in glycogen storage diseases.

At the meeting, Serratrice commented on the different usage of the word by French neurologists. He also expressed concern about the use of different words to denote the same phenomenon as well as using the same word in different languages for different meanings. For instance, "contracture" in English is often used to delineate a permanently shortened muscle, with restriction of active and passive movement around its joints. In French, this condition is called "retraction" or shrinkage [2].

Serratrice [3] has provided a different classification, including: 1) antalgic contracture (as in the spinal tilt and deformity induced by the painful radiculopathy of a herniated intervertebral disc); 2) contracture as the primary cause of pain, as in cramps or the Moersch-Woltman (stiff man) syndrome, or 3) painless contracture in either meaning delineated in chapter 19 by Rowland et al. [1]. In this essay, we will attempt to combine the two approaches.

Antalgic contracture

This term, also called a "compensating contracture" meets the dictionary definition and muscle action potentials are not considered in the three characteristics: 1) it is a secondary response to pain, in which muscles adjacent to the focus contract to assume a posture of least discomfort. The affected muscles are visibly and palpably contracted, standing out from the surrounding muscle mass; 2) the contracture starts from joint; 3) the contracture is transient and disappears, temporarily or permanently, when the pain is relieved and the articulation is released and relaxed. Contrariwise, the contracture is induced or exacerbated by pressure on the painful joint or by mechanical action involving the joint. Electromyographic activity parallels the contracture. For instance, electrical activity in lumbar paraspinal muscles may be seen when the patient sits but the activity disappears with hyperextension of the spine [4].

Perhaps the best known compensating contractures are encountered in lumbar disc disease, resulting in immobility of the affected spinal articulations. Muscle tone relaxes and becomes normal when the spine is hyperextended or the patient lies down, either prone or supine. Sometimes, the contracture becomes chronic, resulting in hyperlordosis. The Lasègue's sign, limitation of straight leg-raising, is attributed to increased pain by sketching affected nerve roots that leads to a transient and localized contracture of paraspinal and posterior thigh muscles. Transient contracture is also seen in painful torticollis or in the bow string sign of ankylosing spondylitis.

Compensating contracture is a nociceptive reflex and is a form of flexor reflex. The reflex loop starts in the affect joint, with associated pain receptors in tendons and aponeurotic insertions. In the polysynaptic pathway, the afferent fibers synapse with alpha motor neurons in the spinal cord. This type of contracture is tightly linked to the affected joint. Nerve impulses pass through the mixed sensory and motor nerve to the posterior roots; then, after polysynaptic activity, the impulse returns to paraspinal muscles through the same branch [5].

Painful contracture

This condition is the opposite of antalgic contracture, where pain is the primary stimulus that results in contracture. Here the muscle contracture comes first and causes the pain. The starting point is not a joint but either the muscle itself or a muscle contraction induced by activity of the central nervous system.

A second difference between antalgic and painful contracture relates to the variability of symptoms. Antalgic contracture is related to position of the limb or spine. Painful contractures are linked more to time, sometimes being paroxysmal.

The most common form of painful contracture is a cramp. Linking the two may lead to confusion, and it might be best to keep cramps in a category separate from any kind of contracture because cramps are brief, lasting seconds or minutes, and also because it has become standard practice to distinguish common cramps (with electromyographic activity) from metabolically induced *contracture* (with no electrical activity) [6]. Common cramps are thought to be neurogenic, with electromyographic bursts of high frequency (200-300 Hz) and high voltage action potentials. Cramps are common in motor neuron disease and probably arise in the perikaryon, although there may also be foci in peripheral nerves. Cramps are also commonly linked to active clinical fasciculation - in the Denny-Brown and Foley syndrome of benign fasciculation and cramps, in motor neuron disease, multifocal motor neuropathy with conduction block, and X-linked bulbospinal muscular atrophy (Kennedy-Alter-Sung syndrome). Cramps may also be seen in diabetic neuropathy, associated with a nocturnal restless legs syndrome [7, 8].

Painful cramps may also be seen in the Woltman-Moersch syndrome (stiff man syndrome), superimposed on chronic overactivity and muscle contraction that leads to abnormal posture [9, 10].

Metabolic *contracture* may also be painful and brief, but the attack is an electrically silent shortening of muscle occuring during exercise, in contrast to the high frequency discharges of a cramp. The clinical features of metabolic contractures are described in chapter 18 [1], including those seen in glycogen storage diseases, Brody's syndrome, and malignant hyperthermia [11].

Painless contracture

This is a heterogeneous group of disorders that includes both myostatic (arising in muscle) and myotaetic (neurogenic). Experimentally, Moll's myostatic contracture is an example [12, 13]. If bones are artificially immobilized, the muscles adjacent to the joint become fixed at a new length. At first, the condition is reversible but, after a few weeks, the number of sarcomeres is reduced permanently. In humans, as discussed in chapters 20 and 22, the fixed and abnormal postures are exemplified by arthrogryposis multiplex congenita, Volkmann ischemic contracture, and the contractures of muscular dystrophies, especially the Emery-Dreifuss and Bethlem varieties. Contractures in muscular dystrophies are attributed to unequal weakness of agonist and antagonist muscles, leading to permanently abnormal postures. That explanation may suffice for Duchenne dystrophy but does not seem to explain the contractures of Emery-Dreifuss, where there may not be much weakness of muscles adjacent to a fixed joint. Myotaetic contracture, seen in states of spasticity, is attributed to an exaggerated monosynaptic spinal reflex.

A suggested glossary

Contracture

Involuntary shortening of one or several muscles, lasting longer than a few minutes.

Myostatic contracture

Equivalent to the French "retraction", permanently fixed and painless shortening of muscle attributed primarily to static immobilization of muscle.

Cramp

A sudden, involuntary and painful shortening of muscle, with visible and palpable knotting of muscle. The EMG shows high frequency discharge.

Metabolic contracture

Shortening of muscle without generation of muscle action potentials (electrically silent); seen in metabolic myopathies.

Myotaetic contracture

Chronic monosynaptic contracture in spastic limbs after a lesion in the central nervous system.

Antalgic contracture

Abnormal posture that arises from neural activity to relieve pain in spine or limb; it may be relieved by change in posture that ameliorates pain.

Spasm

Any sudden, involuntary, brief contraction of a muscle, often due to neural activity, as in facial spasm or blepharospasm.

Rigidity

In English parlance, now restricted to contraction of agonists and antagonists in parkinsonism and other diseases of the basal ganglia, often with tremor ("tremor run together").

Myogelosis, tender strands, trigger points, fibromyositis

Terms not used by neurologists, meant to delineate painful states in muscles that are normal by clinical examination, electromyography, and muscle biopsy. The terms often imply psychogenic disease.

Brachymyoma

Is a rare congenital muscles shortening of unknown mechanism.

References

1. Rowland LP Muscle contracture; physiology and clinical classification. This volume
2. Serratrice G, Simon L (1998) Nosologie, nomenclature, essai de classification des contractures musculaires. In: Simon L et al (eds) Progrès en médecine physique et de réadaptation. Masson, Paris, pp 252-258
3. Serratrice G (1976) Les contractures musculaires rhumatologiques. In: Simon L (ed) Actualites en rééducation fonctionnelle. Masson, Paris, pp 104-112
4. Attali P, Lièvre JA (1965) Recherche de la contracture musculaire chez les sujets atteints de lombalgies et de sciatique. Rev Rhum 32:514-516
5. Lazorthes G (1971) Le système nerveux périphérique. Masson, Paris, vol 1, p 509
6. Rowland LP (1985) Cramps, spasms, and muscle stiffness. Rev Neurol 141:261-273
7. Gorman C, Dyck P, Pearson J (1965) Symptoms of restless legs. Arch Int Med 115:155-160
8. Walters AS, Mickey K, Maltzman J et al (1996) A questionnaire study of 138 patients with restless legs syndrome. Neurology 46:92-95
9. Auger RG (1991) Continuous muscle fiber activity. Semin Neurol 11:285-290
10. Stayer C, Meinck H-M (1998) Stiff-man syndrome: an overview. Neurologia 13:83-8
11. Serratrice G, Kozack-Ribbens G, Cozzone P (1997) Aspects neurologiques de l'hyperthermie maligne. Rev Neurol 153:304-313
12. Détrie Ph (1956) Effets de l'immobilisation sur les articulations normales. Presse Med 64:759
13. Mastromarino R, Maiotti A (1956) Ricerche sperimentali sugli effetti dell' immobilizzazione e sulla repriza/funzionale in articulazion normali. Ortopedia e Traumatologia dell' Apparato Motore 24:173-195

Muscle Contracture: Physiology and Clinical Classification

L.P. Rowland, W. Trojaborg, R.G. Haller

Dorland's medical dictionary defines "contracture" as "*a condition of fixed high resistance to passive stretch of a muscle, resulting from fibrosis of the tissues supporting the muscles or the joints, or from disorders of the muscle fibers*".

O'Dwyer et al. [1] provided a more specific definition in their study of contracture with spasticity after CNS lesions: "*...a shortening of muscle length due to a decrease in the number of sarcomeres in series along the myofibrils, accompanied by an increase in the resistance to passive stretch. Muscle fibers are not lost or replaced by connective tissue...*"

They concluded that contracture arises more from immobility than from spasticity [2].

Table 1. Forms of chronic contracture

Muscular dystrophies	
Post weakness	DMD, BMD
Pre-weakness	EDMD, Bethlem
Congenital myopathies	Fukuyama congenital MD, nemaline, central core; centronuclear
Excessive fibrosis	Bethlem myopathy, myosclerosis
Hyperactivity syndromes	Schwartz-Jampel; Isaacs-Mertens; Stiff-man, Muscle ischemia
Uncertain "primary"	Volkmann, Rigid spine, bent spine syndromes; arthrogryposis
Non-myopathic	CNS lesions with spasticity; post-polio
Idiopathic	Dupuytren, frozen shoulder

Abbreviations: DMD Duchenne Muscular Dystophy; BMD Becker Muscular Dystophy; EDMD Emery-Dreifuss Muscular Dystophy; CNS Central Nervous System; MD Muscular Dystophy

That type can be called "chronic" (Table 1). But there is also another meaning, shortening of a muscle fiber without accompanying muscle action potentials. This "transient" or "pathophysiologic" contracture is seen in three different metabolic conditions: lack of muscle phosphorylase (PPL), phosphofructokinase (PFK) or other enzymes of the glycolytic pathway; malignant hyperthermia; and the Brody syndrome. All three metabolic disorders result from impaired calcium-binding by the sarcoplasmic reticulum, each arising in a special way.

Transient (pathophysiologic) contracture

Heritable block of glycolysis

In his original description of phosphorylase deficiency, Brian McArdle [3] noted that the painful cramps experienced by his patient were electrically silent and he therefore used the word "contracture" to describe it. He also suggested that it might result from impaired glycolytic production of ATP under anaerobic conditions. However, when we tried to prove this theory, we found no consistent loss of ATP during a contracture induced by ischemic work, as measured directly in the affected muscle [4]. We postulated that the time taken for excising the muscle must have allowed time for regeneration of ATP, or that the ATP had been lost in some compartment of the muscle, perhaps the sarcoplasmic reticulum (SR), and that this focal loss was not perceived in measurements of the whole muscle. However, both of those theories were essentially eliminated by studies with magnetic resonance spectroscopy, in which there is no excision of muscle, as summarized by Ruff [5]. He therefore created an animal model by blocking glyceralde-hyde-3-phosphate. In those animals, ischemia induced contracture and the affected muscle showed markedly increased content of ADP and Ca^{2+}, with low values for inorganic phosphate, and failure to acidify. This combination of metabolic aberrations increased the sensitivity of contractile proteins to calcium in the presence of an excessive content of calcium ions, disrupting the "interplay among the contractile proteins, calcium release, and calcium sequestration", all of which could lead to contracture at first, then to necrosis and myoglobinuria. Haller et al. [6] found lower levels of Na^+ - K^+ ATPase, which would impede potassium reuptake, further distorting the normal sarcolemmal environment.

Brody syndrome

This disorder is more fully described by Dr. Karpati [7], but it is noteworthy how complete Brody's description was [8]. He noted all of the essential findings: the slowness of relaxation without electrical activity, as demonstrated by combined electromyography and tension measurements, and impairment of calcium binding by preparations of sarcoplasmic reticulum. Karpati et al. [9] confirmed the findings almost 20 years later, identifying the defective enzyme as the Ca^{2+} - Mg^{2+} ATPase of the SR. At last count, 22 patients had been described, 10 in one report [10]. One patient had attacks of necrosis with myoglobinuria [11]. SERCA1 is the isoform of the calcium-binding ATPase in the SR of fast-twitch muscle; was the first candidate gene and is affected in some families [10] but has been eliminated in other families [7, 12], providing evidence of locus heterogeneity.

We have seen only one patient. At age 23, she had had typical symptoms and Trojaborg [13] described the results as follows: "During tetanic stimulation of the ulnar nerve at 20 Hz for 8 sec with simultaneous recording of the mechanical force and the electrical muscle response from the abductor pollicis, the developed force was 5 Newtons, and the amplitude of the compound muscle action potential was 9mV throughout. However, the half-relaxation time at the end of the stimulation was 240 msec, 2.5 times normal. The time from the last action potential to a return of muscle force to zero after tetanic stimulation was 1.080 msec, four times longer than normal. At a low rate of stimulation (2 Hz in 100 sec), the duration of the mechanical responses increased significantly during continuous stimulation, contrary to findings in normal subjects in whom the responses become shorter with time". The findings were deemed characteristic of Brody disease, and were sufficient to exclude the possibility that the slowness of relaxation might be volitional.

Haller [14] reported the following: "We found no distinctive metabolic abnormalities to correlate with the muscle stiffness and slowed muscle relaxation she experiences in exercise. During cycle testing, exercise was limited by symptoms of stiffness and aching of exercising muscle (quadriceps and gastrocnemius) associated with a slowing of pedal rate. The peak increases in heart rate and in blood lactate and pyruvate were attenuated, consistent with submaximal symptom-limited exercise. During ischemic and aerobic forearm testing, lactate and ammonia production were within normal limits. The rate of decline in contractile force was within broad limits of normal, but the rate of muscle relaxation clearly slowed to a greater than normal degree as indicated by the changes in the force tracing in the last half of this test. ^{31}Phosphorus magnetic resonance spectroscopy over the anterior tibial muscle at rest, during exercise, and with recovery was within normal limits. There was an approximate doubling of serum creatine kinase from 200 to 400, a bit more than usually seen with this level of exercise but not definitely abnormal".

We have not published these findings previously because the patient has not permitted a muscle biopsy to prove the diagnosis by directly measuring ATPase activity and calcium uptake.

Malignant Hyperthermia (MH)

This condition is defined clinically [15], with the following elements: exposure to both halothane and succinylcholine choline (or similar agents); sharp rise in body temperature; stiffness of masseter and limb muscles (presumed to be contracture); lactic acidosis; cardiac arrhythmia; muscle necrosis and sky high values for creatine kinase with myoglobinuria. When the syndrome was first recognized the fatality rate exceeded 50% but that has been lowered by dantrolene therapy.

MH is attributed to an energy crisis in muscle. Much of the evidence has been taken from studies of a strain of susceptible pigs, as described in detail by Mickelson and Louis [16]. The genetic fault is attributed to the ryanodine receptor (RYR1), a calcium-release channel in the SR. As a result, sensitivity to depolarizing muscle blockers and anesthetics is exaggerated, the regulation of cytosolic Ca^{2+} goes away, the muscle cell is flooded with calcium, and excitation-contraction coupling is distorted as troponin C is activated, accelerating the interaction of actin and myosin to generate contractile force. Actomyosin ATPase activity depletes muscle stores of ATP, initiating resynthesis of ATP by increasing the rates of both aerobic and anaerobic metabolism, and disaster ensues.

Human MH susceptibility has been mapped to the same gene for RYR1 at 19q13.1 but mutations here account for less than 10% of susceptible families. Locus heterogeneity is therefore evident, but other genes have not been identified with certainty. Central core disease has also been mapped to 19q13.1 and patients with that myopathy seem to be at increased risk for MH.

Part of the problem in finding the genes is that of identifying MH "susceptibility" because the best biological marker has been the *in vitro* contracture test, a measure of exaggerated sensitivity of muscle strips to caffeine. The test is thought to be generally accurate but not 100% sensitive or specific, so that results are sometimes falsely negative or positive.

The neuroleptic malignant syndrome shares many clinical features with MH but is thought to have a different pathogenesis [17, 18]. The neuroleptic disorder is attributed to abrupt blockade of central dopaminergic receptors. Survivors of the neuroleptic syndrome are not thought to be at increased risk of MH. The muscle "rigidity" of neuroleptic sensitivity presumably arises from activity of the central nervous system and would therefore not be "contracture".

Comparing pathophysiologic contracture mechanisms

All three transient contracture syndromes result from excessive Ca^{2+} content in muscle, but the mechanisms differ. The glycogen diseases cause symptoms because glycolysis is blocked and energy metabolism is altered in a complex manner. MH and Brody syndrome both arise from primary abnormalities of calcium-binding by the sarcoplasmic reticulum, but one is attributed to the ryanodine receptor, a calcium releasing channel, and the other is attributed to an altered ATPase of the SR. All three mechanisms have been difficult to pin down.

Chronic contracture

Contractures after immobility

There seems to be little information about the pathogenesis of limb or spine contractures. It is often assumed that lack of motion about a joint will result in fibrosis around the joint and then to fixed abnormal postures; it is also assumed that this can be ameliorated or prevented by appropriate preventive physical therapy with assisted passive motion of the totally immobile articulation, and assisted active exercise when possible. However, O'Dwyer et al. [1, 2] believe there are changes in immobilized muscles that resist passive motion. This pattern is seen in spastic limbs after cerebral or spinal cord injury [1, 2], or after poliomyelitis. A similar mechanism may account for the contractures seen in Duchenne and Becker dystrophy, which may be prevented or delayed by prophylactic exercises. However, it is then not clear why contractures of that degree are less a problem in spinal muscular atrophy, myotonic muscular dystrophy, or facioscapulohumeral muscular dystrophy. These disorders are described by Kakulas [19] and Fardeau [20].

Contractures are encountered in the congenital muscular dystrophy of Fukuyama and also in some structurally specific congenital myopathies. One form of congenital myasthenia gravis was also associated with contractures [21]. In these conditions it is not clear whether the joint disorders are secondary to muscle weakness or are established independently.

Early-onset contractures (arthrogryposis, Emery-Dreifuss syndrome, Bethlem myopathy, rigid spine syndrome, bent spine syndrome)

In contrast to the contractures of immobility, some syndromes are characterized by early onset of contracture before there is much, if any, detectable local weakness. This pattern is characteristic of arthrogryposis congenita (Munsat, this volume) and it is defined in the X-linked recessive and autosomal dominant forms of Emery-Dreifuss muscular dystrophies, Bethlem autosomal dominant muscular dystrophy, laminin-deficient autosomal dominant limb girdle muscular dystrophy and variants of the rigid spine disease. As a result of the prominence of contractures in these disorders, there is much clinical overlap and diagnostic confusion, which makes positional cloning the more difficult.

Emery-Dreifuss muscular dystrophy is usually X-linked recessive and is defined by the clinical triad of (1) early contractures at the elbows, Achilles tendons, and neck extensors; (2) myopathic humeroperoneal muscle wasting and weakness, and (3) heart block [22-24; Arahata, this volume]. The affected gene product is emerin, which has been identified as a nuclear protein; it is expressed so ubiquitously that the diagnosis can be made by skin biopsy [25]. How that might lead to contractures, however, is not clear. The same clinical features are found in the autosomal dominant form [22, 26], which has not yet been mapped.

Bethlem myopathy resembles Emery-Dreifuss in the location of early contractures and also in the anatomic distribution of weakness. It differs in that the cardiopathy is absent and inheritance is autosomal dominant. Dutch families were linked to 21q22.3, the site of the gene for type VI collagen, which could be a clue to the prominence of the contractures [27]. Locus heterogeneity has been evident because French Canadian families linked to 2q37 [28].

The rigid spine syndrome has long been regarded as one of diverse etiology because it has been associated with diverse patterns of clinical and muscle pathology. But the spinal disorder can be seen in Emery-Dreifuss disease, and there may be joint contractures in the rigid spine syndrome. Lotz and Stubgen [29] described a consistent pattern of vacuolar change in muscle, but Lotz et al [1997] later found that vacuoles might be indicative of late-onset acid maltase deficiency. Reichman et al. [30] found features of cytoplasmic bodies, desmin accumulation, reducing bodies, and tubulofilamentous inclusions, all in one patient.

The obverse of the rigid spine is the "bent spine", in which the dominant symptom is inability to stand erect; the syndrome was described by Serratrice et al. [31]. Katz et al. [32] relate a similar syndrome of "isolated neck extensor myopathy" to the "floppy head" [33] or "dropped head" [34] syndromes, but in the only patient I have personally examined, the outstanding finding was the patient's inability to stand erect ("camptocormia"). His chief complaint was: "I look like Quasimodo". His posture resembled the Moersch-Woltman or stiff-man syndrome. One hallmark noted by Serratrice and his colleagues was the replacement of paraspinal muscles by fat, as seen clearly by computerized tomography. They attributed the inability to stand erect to weakness of paraspinal muscles rather than contracture because the abnormal posture disappeared when the patients were supine.

Another unique disorder with fixed joints in abnormal postures is "myosclerosis", which was described by Bradley et al. [35]. They thought it might be a consequence of polymyositis and their patient improved with penicillamine therapy. Their paper was followed by an almost total lack of later reports [36]. I have seen one patient who might qualify, but formal studies did not show any unique histopathology.

At age 23, this woman delivered her first child. In the process she found that she could not supinate the left arm for placement of an intravenous infusion. At age 25, she could not abduct her arms in an exercise program and two years later examination showed she could not raise her arms overhead. There were contractures at the elbows and shoulders. Muscles of the left shoulder were firm and hard to palpation. The left elbow was flexed and abducted. There was no weakness on manual testing. Sclerodactyly was noted and she recognized Raynaud symptoms. Muscle biopsy, however, showed mostly normal muscle; a small amount of fibrosis could have been normal epimysium at the edge of the muscle (Dr. A. P. Hays). EMG findings were consistent with myopathy. At age 34, there had been no change and MRI showed no abnormality of the shoulder girdle muscles. The following tests gave normal results: antinuclear antibodies, rheumatoid factor, immunofixation electrophoresis, and plain roentgenogram for calcification in the hands. There was no family history of similar disease.

Schwartz-Jampel syndrome

This is another disorder characterized by early onset of contractures in the absence of much muscle weakness. It is defined by the clinical features: (1) musculoskeletal abnormalities, with an asymmetric head, blepharophimosis, pursed lips, short stature and short limbs, bony abnormalities and multiple limb contractures, and (2) slow relaxation after a forceful contraction, clinically resembling myotonia and sometimes accompanied by true myotonic discharges in the electromyogram but also sometimes with continuous motor unit activity, myokymia, and complex high frequency discharges of "neuromyotonia" [37, 38]. Taylor et al. [37] found continuous motor unit activity, with no electrical silence in apparently resting muscle; they therefore likened the disorder to the Isaacs-Mertens syndrome (neuromyotonia). They found that the discharges did not wax and wane as in true myotonia and that curare abolished the Schwartz-Jampel activity, although it does not affect true myotonia, which is therefore attributed to a muscle membrane disorder. However, Taylor et al also found that the spontaneous activity persisted after a paralytic nerve injury, and some other investigators [39, 40] found curare ineffective in relieving the continuous activity. These observations suggest that there may sometimes be an inherent muscle abnormality, that the syndrome may involve both muscle and nerve - not solely one or the other [38].

But it is not clear why contractures should be so prominent in the Schwartz-Jampel syndrome. The disorder it most closely resembles physiologically is the Isaacs-Mertens syndrome (neuromyotonia), in which continuous muscle fiber activity is also prominent and thought to arise in peripheral nerves. In that disorder, there may be abnormal postures of the hands and feet, resembling carpopedal spasm, but limb contractures are rarely prominent [41]. Why then are the abnormal limb contractures so prominent in Schwartz-Jampel?

At least three forms of Schwartz-Jampel syndrome are recognized. The most common is autosomal recessive; symptoms begin in late infancy or childhood and it has been mapped to chromosome 1p34-p36.1 [42]. A second, more severe, type is the neonatal variety which may be fatal and which does not map to chromosome 1 [43, 44]. We have recently seen one such case. There was no family history of similar disease. The congenital form of myotonic muscular dystrophy was considered in the differential diagnosis but the mother showed no signs of that disease and DNA analysis for the infant did not conform to the expanded repeat of MyD. The third type is autosomal dominant [38]. The origin of the contractures in these conditions is not known.

An overall view

Chronic contracture is the source of much disability but it has somehow escaped modern approaches to the pathophysiology. This conference may direct much needed attention to analysis of the problem. If so, it may enhance the management of people with myopathies, CNS lesions, and hyperactivity syndromes of the motor unit. We know more about the pathophysiology of transient contractures, but even there we have much to learn about the origin and treatment of the causal conditions.

Acknowledgement

Professors W. Trojaborg and R.G. Haller provided the data presented here from their studies of a patient with the Brody syndrome. Dr. Arthur P. Hays has always been helpful in the interpretation of muscle biopsies, including the case of possible myosclerosis described here. Ms. Sheila Crecesenzo and Mr. Terrance Gabriel provided invaluable assistance in preparing the lecture and manuscript. Dr. Dora Leung was helpful in dealing with patients and reviewing the manuscript.

References

1. O'Dwyer NJ, Ada L (1996) Reflex hyperexcitability and muscle contracture in relation to spastic hypertonia. Curr Op Neurol 9:451-455
2. O'Dwyer NJ, Ada L, Neilson PD (1996) Spasticity and muscle contracture following stroke. Brain 119:1737-1749.
3. McArdle B (1951) Myopathy due to defect in muscle glycogen breakdown. Clin Sci 10:13-35
4. Rowland LP, Araki S, Carmel P (1965) Contracture in McArdle's disease. Stability of adenosine triphosphate during contracture in phosphorylase-deficient human muscle. Arch Neurol 13:541-544
5. Ruff RL (1988) Why do patients with McArdle's disease have decreased exercise capacity? Neurology 50:6-7

6. Haller RG, Clausen T, Vissing J (1998) Reduced levels of skeletal muscle Na^+-K^+-ATPase in McArdle disease. Neurology 50:37-40
7. Karpati G Brody's syndrome. This volume
8. Brody IA (1969) Muscle contracture induced by exercise. A syndrome attributable to decreased relaxing factor. N Engl J Med 281:187-191
9. Karpati G, Charuk J, Carpenter S, Jablecki C, Holland P (1986) Myopathy caused by a deficiency of Ca^{2+} adenosine triphosphatase in sarcoplasmic reticulum (Brody's disease). Ann Neurol 20:38-49
10. Benders AAGM, Veerkamp JH, Oosterhof A et al (1994) $Ca2^+$ homeostasis in Brody's disease. A study in skeletal muscle and cultured muscle cells and the effects of dantrolene and verapamil. J Clin Invest 94:741-748
11. Poels PJE, Wevers RA, Breasskhekke JP, Benders AAGM, Veerkamp JH, Joosten EMG (1993) Exertional rhabdomyolysis in a patient with calcium adenosine triphosphatase deficiency. J Neurol Neurosurg Psychiatry 56:823-826
12. Zhang Y, Fujii J, Phillips MS, Chen HS, Karpati G, Yee WC, Schrank B, Cornblath DR, Boylan KB, Maclennan DH (1995) Characterization of cDNA and genomic DNA encoding SERCA1, the Ca^{2+}-ATPase of human fast-twitch skeletal muscle sarcoplasmic reticulum and its elimination as a candidate gene for Brody's disease. Genomics 30:415-424
13. Trojaborg W (1994) Personal communication, written report
14. Haller R (1997) Personal communication, written report
15. Tein I, DiMauro S, Rowland LP (1992) Myoglobinuria. In: Rowland LP, DiMauro S (eds) Myopathies. Hdbk Clin Neurol, Elsevier Science Publishers, Amsterdam, vol 18, pp 553-593.
16. Mickelson JR, Louis CF (1996) Malignant hyperthermia: excitation-contraction coupling, Ca^{2+} release channel, and cell Ca^{2+} regulation defects. Physiol Rev 76: 538-592
17. Keck PE Jr, Caroff SN, McElroy SL (1995) Neuroleptic malignant syndrome and malignant hyperthermia: end of a controversy? J Neuropsychiatry Clin Neurosci 7:135-144
18. Bertorini TE (1997) Myoglobinuria, malignant hyperthermia, neuroleptic malignant syndrome and serotonin syndrome. Neurol Clin 15:649-671
19. Kakulas BA Pathologic aspects of muscle contracture. This volume
20. Fardeau M Phenotypic and genetic heterogeneity of muscle contractures. This volume
21. Smit LME, Jennekens FGI, Veldman H, Barth PG (1984) Paucity of secondary synaptic clefts in a case of congenital myasthenia gravis with multiple contractures; ultrastructural morphology of a developmental disorder. J Neurol Neurosurg Psychiatry 47:1091-1097
22. Emery AEH (1989) Emery-Dreifuss syndrome. J Med Genet 26:637-641
23. Rowland LP, Fetell M, Olarte M, Hays A, Singh N, Wanar FE (1979) Emery-Dreifuss muscular dystrophy. Ann Neurol 5:111-117
24. Tsuchiya Y, Arahata K (1997) Emery-Dreifuss syndrome. Curr Op Neurol 10:421-425
25. Mora M, Cartegni L, Bi Blasi C et al (1997) X-linked Emery-Dreifuss muscular dystrophy can be diagnosed from skin biopsy or blood sample. Ann Neurol 42:249-53
26. Yates JRW (1997) European workshop on Emery-Dreifuss muscular dystrophy. Neuromusc Disord 7:67-69
27. Jobsis GJ, Keizers H, Vreijling JP et al (1996) Type VI collagen mutations in Bethlem myopathy, an autosomal dominant myopathy with contractures. Nature Genet 14:113-115
28. Speer MC, Tandan R, Rao PN et al (1996) Evidence for locus heterogeneity in Bethlem myopathy and linkage to 2q37. Hum Mol Genet 5:1043-1046
29. Lotz BP, Stubgen JP (1993) The rigid spine syndrome: a vacuolar variant. Muscle Nerve 16:530-536
30. Reichmann H, Goebel HH, Schneider C, Toyka KV (1997) Familial mixed congenital myopathy with rigid spine phenotype. Muscle Nerve 20:411-417

31. Serratrice G, Pouget J, Pellissier JF (1996) Bent spine syndrome. J Neurol Neurosurg Psychiatry 60: 51-54
32. Katz JS, Wolfe GI, Burns DK, Bryan WW, Fleckenstein JL, Barohn RJ (1996) Isolated neck extensor myopathy; a common cause of dropped head syndrome. Neurology 46:917-921
33. Lange DJ, Fetell MR, Lovelace RE, Rowland LP (1986) The floppy head syndrome (abstract). Ann Neurol 20:133
34. Suarez GA, Kelly JJ (1992) The dropped head syndrome. Neurology 42:1625-1627
35. Bradley WG, Hudgson P, Gardner-Medwin D, Walton JN (1973) The syndrome of myosclerosis. J Neurol Neurosurg Psychiatry 36:651-660
36. Ricker K, Seitz D, Trostdorf E (1970) Myositis fibrosa generalisata and "stiff man syndrome". Europ Neurol 11:13-27
37. Taylor RG, Layzer RB, Davis HS, Fowler WM (1972) Continuous muscle fiber activity in the Schwartz-Jampel syndrome. EEG Clin Neurophysiol 33:497-509
38. Pascuzzi RM, Gratianne R, Azzarelli B, Kincaid JC (1990) Schwartz-Jampel syndrome with dominant inheritance. Muscle Nerve 13:1152-1163
39. Cao A, Glanchetti C, Calisi L, DeVirgilis S, Ferreli A, Tangheroni W (1978) Schwartz-Jampel syndrome: clinical, electrophysiological, and histopathological study of a severe variant. J Neurol Sci 15:5-12
40. Spaans F, Theunissen P, Reekers AD, Smit L, Veldman H (1990) Schwartz-Jampel syndrome: I. Clinical, electromyographic, and histologic studies. Muscle Nerve 13:516-527
41. McGuire SA, Tomasovic JJ, Ackerman N Jr (1984) Hereditary continuous muscle fiber activity. Arch Neurol 41:395-396
42. Nicole S, Ben Hamida C, Beighton P et al (1995) Localization of the Schwartz-Jampel syndrome locus to chromosome 1p34-36.1 by homozygosity mapping Hum Mol Genet 4:1633-1636
43. Al-Gazali LI, Varghese M, Varady E, Al Talabani J, Scoerer J, Bakalinova D (1996) Neonatal Schwartz-Jampel syndrome; a common autosomal recessive syndrome in the United Arab Emirates. J Med Genet 33:203-211
44. Brown KA, Al-Gazali LI, Moynihan M, Lench NJ, Markham AF, Mueller RF (1997) Genetic heterogeneity in Schwartz-Jampel syndrome: two families with neonatal Schwartz-Jampel syndrome do not map to chromosome 1p34-p36.1. J Med Genet 34:685-687
45. Fadic R, Wacwik AJ, Brooks BR, Lotz B (1997) The rigid spine syndrome due to acid maltase deficiency. Muscle Nerve 20:364-366
46. Isaacs H, Badenhorst ME (1993) False negative results with muscle caffeine halothane contracture testing for malignant hyperthermia. Anesthesiology 79:1-4
47. Larach MG, Landis Jr (1990) False positive diagnosis of MH susceptibility on control subjects using the North American caffeine halothane contracture test. Anesthesiology 73:A1013
48. Rowland LP (1985) Cramps, spasms, and muscle stiffness. Rev Neurol (Paris) 141:261-273
49. Taylor J, Muntoni F, Robb S, Dubowitz V, Sewry C (1997) Early onset autosomal dominant myopathy with rigidity of the spine; a possible role for laminin b1? Neuromusc Disord 7:211-216

Pathologic Aspects of Muscle Contracture

B.A. KAKULAS

Contractures of muscle are a major cause of disability in muscular dystrophy of the Duchenne type and to a lesser degree in other neuromuscular disorders. Limitation of joint movement contributes to the patients disability because of the resultant aggravation of immobility. In Duchenne muscular dystrophy (DMD) joint and muscle contractures develop as the disease progresses. The same applies for some other forms of muscular dystrophy. There is also a group of neuromuscular disorders where the joint contracture is present at birth, chief among these is arthrogryposis. In others, muscle contractures develop soon after birth and are among the first manifestations of the disorder, e.g. the Emery Dreifuss form of muscular dystrophy. Contractures are also common in the neurogenic diseases, notably SMA II, the chronic infantile form of Werdnig-Hoffman disease, and in the post polio syndrome. Contractures also occur in acquired myopathies such as polymyositis.

In general terms, muscle contracture is of two types, pathophysiological and structural (Table 1). The pathophysiological forms are localised muscle contractions such as in cramps and the electrically silent contractures of McArdle's disease. Generalised myostatic contracture is secondary to immobilisation in which the overall length of the muscle fibre is reduced. Structural contractures are due to fibrosis of muscle and ankyloses of joints.

Classification and definition

Cramps

'Electrical' cramps occur in normal individuals and as a symptom in many neuromuscular diseases. There is also a familial cramping syndrome [1]. Cramps are common in DMD and in the early stages of amyotrophic lateral sclerosis (ALS). 'Electrical' muscle cramps are due to volleys of depolarisation transmitted from the motor nerves causing intense and usually painful contractions of many motor units. The gastrocnemius is particularly prone. Characteristically cramps may be reversed by the simple extension of the muscle, in the case of the gastrocnemius, dorsiflexion of the foot is often effective. The cause of the discharge volley in electrical cramps is unknown but is believed more likely to be due to dehydration than to electrolyte deficiencies [2]. In DMD cramps may be due to the entry of calcium into the cytoplasm due to the absence of dystrophin causing minor disruption of the muscle membrane.

Table 1. Muscle Contracture

• Muscular cramps
• Electrically silent contractures
• Myostatic contracture
• Fibrous contracture
• Arthrogryposis
• Ankylosis

Muscle contracture

Localised contractures occur as a result of some inherent abnormality within the muscle, usually of metabolic origin and are typical of McArdle's disease caused by myophosphorylase deficiency or Tarui's disease due to phosphofructokinase deficiency [3]. Muscle contractures are silent and usually very painful, although painless contractures are also known. Classic electrical cramps may also occur in these disorders of glycogen metabolism.

Myostatic contracture

In myostatic contracture the overall length of the muscle is reduced and the joint fixed in flexion. It is well established that the length of a particular muscle and its constituent muscle fibres is determined by the degree of extension applied to the muscle on a daily basis. There is a very powerful feedback loop which controls the length of the muscle fibre according to this principle. If a limb is placed in a plaster cast and the joint immobilised, after a few weeks it is found that it is not possible to fully extend the joint. This is due to actual shortening of the muscle and each individual muscle fibre *per se*. Myostatic contractures are not due to fibrosis of the muscle tendon or joint but simply reflect the reduced range of movement at the joint [3]. Prolonged immobilisation of any joint will lead to myostatic contractures and is a common complication of fixed deformities and plaster casts but are most often encountered in the muscular dystrophies and some congenital myopathies. Extreme rigidity of the spine presumably due to fibrous and myostatic contracture is a feature of several congenital myopathies especially the minicore type. Myostatic contractures are amenable to treatment by physical means. The therapist stretches the muscle to its full length at least once daily for up to 30 minutes. As a result of this stretching stimulus, individual muscle fibers gradually lengthen by the addition of new sarcomeres [3].

Fibrous contracture

This term refers to the fibrous contractures which may be encountered in a great variety of neuromuscular disorders, especially at their end stage. In the severe forms of muscular dystrophy, mainly represented by DMD and similar phenotypes within the sarcoglycanopathy group, there is progressive destruction of muscle fibres with their replacement by fat and fibrous tissue [4].

Arthrogryposis and ankylosis

Arthrogryposis refers to limitation of joint movement present at birth of which there are several forms. It is usually secondary to reduced mobility in utero from a variety of causes. Ankylosis is due to joint disease and occurs in all forms of arthritis and is related to inflammation and fibrosis of the capsule and ligaments.

Pathological aspects of contractures

Cramps

Usually no morphologic abnormality can be found in muscles prone to physiological (electrical) cramps. If the cramps are frequent as occurs in the familial cramping syndrome, hypertrophy of individual muscle fibres may be present. Ringbinden are also said to be occasionally present, but these are more often associated with the myotonic disorders, especially Thomsen's disease [5].

Muscle contractures

The silent contractures of McArdle's disease and related metabolic disorders are usually not associated with morphological changes except for those which are due to the primary disease in the form of excess intra-fibre glycogen. Enzyme histochemistry reveals an absence of myophosphorylase in McArdle's disease and phosphofructokinase in Tarui's disease.

Myostatic contracture

Biopsy of a muscle which has contracted and shortened due to prolonged immobilisation usually shows little microscopic change except for atrophy, especially of the histochemical type 2 muscle fibres. In myostatic contracture, new sarcomeres will be laid down following passive stretching which is used as a treatment by physical therapists. The muscle in this case is subject to regular stretch, sufficiently firm to stimulate the desired lengthening but not so great as to cause disruption and haemorrhage [6].

It is said, but not adequately demonstrated, that collagen may realign in myostatic contracture and thus be responsible for fibrous shortening. The rearrangement of collagen orientation and increase in quantity of fibrous tissue in myostatic contracture remains unproven.

Fibrous contractures

The myopathology of the more severe forms of muscular dystrophy, especially DMD, is characterised by increased quantities of fibrous tissue as well as fat. It is believed that this is a condensation and simply a reflection of the loss of muscle fibres and not a true proliferation. However, it is evident that even in the early stages of DMD, the endomysial fibrous tissue is more obvious and thick. It is possible that the primary molecular defect resulting from the absence of dystrophin in some way stimulates collagen deposition. Nevertheless, the main principle is the inherent property of connective tissue to contract, i.e. shorten with the passage of time. The main pathologic finding in the muscular dystrophies with fibrous contractures is to find an excess of connective tissue within the affected muscles.

Collagen has the property of shortening which is a universal effect in all scars. This matter has been studied by Kottke et al. [7] who points out that normal connective tissue develops a tensile force which causes a slow progressive shortening of connective tissue until stopped by an opposing force. This property of progressive shortening is well illustrated in the muscular dystrophies as a result of chronic imbalance across joints and is aggravated by sitting for long periods with flexed hips and knees. The connective tissue gradually contracts with reorganisation of collagen to produce dense connective tissue and permanent restriction of motion. Fibrous contracture is a feature of end stage polymyositis and the neurogenic atrophies, especially poliomyelitis, as well as occurring in DMD and related dystrophic conditions.

Arthrogryposis and ankylosis

Arthrogryposis due to muscle contractures which have developed *in utero* may be due to a great variety of causes. Any condition which limits movement *in utero* such as primary muscle disease, peripheral neuropathy or agenesis of motor neurons and local joint diseases may all be responsible (Table 2). In an analysis of 30 of our cases, a neuropathic cause was undoubtedly present in 7 cases, 2 with anterior horn cell disease and 5 with a neuropathic picture with peripheral nerve involvement and quite possibly involvement of the anterior horn cells as well. Three patients had undoubted primary myopathies, 2 with myotonic dystrophy and 1 with congenital myotubular myopathy. Seven had major cerebral abnormalities with no definite abnormalities in nerve and muscle in 3 and insufficient data in 4. Two patients with myelodysplasia without spina bifida had muscle abnormalities of which one was associated with sacral agenesis. Three patients would fit the description of amyoplasia. Four patients conform to the diagnosis of distal arthrogryposis and one patient had clasped thumbs due to absence of the extensor tendons. Three had marked oligohydramnios with leaking liquor for many weeks of whom one had sclerema like lesions, one had features of the Pena Shokeir Syndrome and one had features of distal arthrogryposis type 2 with additional abnormalities and tendency to apnoeic spells and oligophrenia [8].

Table 2. Arthrogryposis in western Australia

• Neuropathic (2 AHC, 5 PN)	7
• Primary myopathy (DM and myotubular)	3
• Cerebral abnormalities	7
• Myelodysplasia	2
• Amyoplasia	3
• Distal arthrogryposis	4
• Oligohydramnios	3
• Pena Shokeir syndrome	1
TOTAL	**30**

Ankylosis is a manifestation of joint disease. The inflammatory forms, especially rheumatoid arthritis, are associated with pannus formation, fibrous adhesions and deformity of joints, all of which limit movement. The other main form of arthritis is the degenerative osteoarthrosis wear and tear type which causes limitation of joint movement with the production of osteophytes secondary to the destruction of articular cartilage. Progressive systemic sclerosis and the many other connective tissue diseases associated with arthritis are also causes of ankylosis.

Discussion

Muscle contracture is a predominant symptom and sign in many forms of neuromuscular disease. Patients complain of muscle pain or limitation of movement and discomfort as a result of contractures and their complications. A particularly insidious complication is scoliosis due to weakness and fibrosis of the paraspinal muscles as it occurs in several of the congenital myopathies as well as in Duchenne muscular dystrophy.

Similar comments apply to other forms of neuromuscular disease, especially the spinal muscular atrophies and poliomyelitis, in which myostatic contracture and fibrous contractures are common. Muscle contractures may also be a prominent feature of polymyositis and systemic sclerosis because of the proliferation of connective tissue in association with the chronic inflammatory process. In this case, shortening of the connective tissue may result in the rapid development of contractures so that it becomes one of the important features requiring treatment by the physical therapist.

Various forms of muscle contracture have been considered, each occurring in a particular context. The more important of these is myostatic contracture and scoliosis as they occur in DMD because of their detrimental effect on mobility and respiratory function. A first principle in the rehabilitation of boys with DMD is to maintain a full range of movements for each joint particularly the lower limbs. This is achieved by intensive daily physical therapy according to methods which have been developed over many years.

Because of their prominence within the clinical features of the neuromuscular disorders, but especially the muscular dystrophies, a knowledge of the pathological basis of each form of contractures is of value to the clinician and allied health professionals who are responsible for management.

Summary

Contractures of muscle are a major cause of disability in muscular dystrophy of the Duchenne type and to a lesser degree in the other forms. Limitation of joint movement contributes to the patients impairment because of the postural deformities and aggravation of immobility. In Duchenne muscular dystrophy (DMD) joint and muscle contractures develop as the disease progresses. Muscle contractures are among the first manifestations of several other neuromuscular disorders, e.g. the Emery-Dreifuss form of muscular dystrophy. There is also a group of neuromuscular disorders in which contractures are present at birth, chief among these is arthrogryposis. Contractures are also common in the hereditary spinal muscular atrophies notably SMA II, the chronic infantile form of Werdnig-Hoffman disease and other neurogenic conditions such as the post-polio syndrome. Contractures also occur in other acquired disorders, e.g. chronic polymyositis. In general terms, muscle contracture is classified into two broad groups. Firstly, those which have a functional or physiological basis and secondly, those which are due to structural lesions. The functional forms are the localised muscle contractions such as true (electrical) cramps and the silent contractures of McArdle's disease. Myostatic contracture in which the muscle *per se* is shortened is secondary to immobilisation of joints. Another structural type is the fibrous contracture of muscle which occurs at the end stages of both muscular dystrophy and neurogenic atrophy. When viewed with the microscope, no visible lesions are associated with the physiological forms of contracture except perhaps for an occasional ringbinden and moderate muscle hypertrophy. In myostatic contracture, the initial change is a shortening of individual muscle fibres due to adsorption of sarcomeres. Myostatic contractures are a special problem in DMD where imbalance of opposing muscles, e.g. weak anterior tibial muscles versus the stronger gastrocnemius and soleus, is an aggravating factor. Scoliosis with contracture of paraspinal muscles is a serious complication of DMD. In the end stages of all of the muscular dystrophies shortening of a whole muscle is due to the shrinkage of fibrous tissue which has accumulated as the result of the ongoing myopathic process. Similarly, in the late stages of the neurogenic atrophies fibrous contracture occurs. In chronic polymyositis and progressive systemic sclerosis there is a proliferation of fibrous tissue which is subject to shortening. Ankylosis of joints in arthritis is another cause of myostatic contracture. Contractures in arthrogryposis are due to a variety of myopathic or neurogenic causes related to foetal immobility *in utero.*

References

1. Ricker K, Moxley RT (1990) Autosomal dominant cramping disease, Arch Neurol 47:810-812
2. Griggs RC, Mendell JR, Miller RG (1995) Evaluation and Treatment of Myopathies. FA Davis Company, Philadelphia
3. Kakulas BA, Adams RD (1985) Diseases of muscle: Pathological Foundations of Clinical Myology. 4th edn, Harper and Row Publishers, Philadelphia
4. Kakulas BA (1997) Problems and potential for gene therapy in Duchenne muscular dystrophy. Neuromuscular Disorders 7:319-324
5. Dubowitz V (1985) Muscle Biopsy, A Practical Approach. 2nd edn, Baillière Tindall, London
6. Vignos PJ (1979) Rehabilitation in the Myopathies. In: Vinken PJ, Bruyn GW (eds) Handbook of Clinical Neurology. Diseases of Muscle Part II. Elsevier North-Holland Publishing Co, Amsterdam, chap 13, pp 457-500
7. Kottke FJ, Pauley DL, Ptak RA (1966) The rationale for prolonged stretching for correction of shortening of connective tissue. Arch Phys Med 47:345-352
8. Silberstein EP Personal Communication

The Enigma of Muscle Contractures

M. Fardeau

The first difficulty in this topic is a semantic one. It is generally accepted that "contracture" means a resistance to passive stretch of the muscle; in that way, contracture is clearly different from an impairment of muscle relaxation, such as in myotonias. But it is not precised in which physiological status is the muscle when it resists to the stretch. Is it in an active state of contraction, or not, "contracted" and not active, etc. Curiously, English and French languages are in disagreement about that. In English, the word "contracture" is mainly - but not always - limited to the designation of a permanent shortened position, without any physiological evidence of activity. When the muscle is stimulated, the terms of spasticity, hypertonicity, spasms, stiffness and rigidity are used to characterize the status of the muscle, whatever was the duration of the major resistance to a passive elongation - as, for instance, - in cerebral palsy. In French, it is the opposite : the word "contracture" is used to designate a status of intense, often painful, contraction of the muscle, as in tetanus, dystonia, neuromyotonia, or even hysteria ("contracture hystérique"). And, when the muscle shows no sign of activity, French authors do not use the word contracture, but the word "rétractions" (Table 1).

It should be emphasized that an intermediate zone between the English and French terminologies is dealing with muscles contracted, but electrically silent, like in paramyotonia, Brody's disease or myoedema, as well as in McArdle's disease. Most of the time, in the English textbooks, these muscle conditions are designated as post-exercise "cramps", cramping, or stiffness, but sometimes the word "contracture" is used [1, 2]. In French, the word "contracture" is commonly used for these conditions. Anyway, we shall only consider in this paper the English meaning of the term "contracture", as observed during the course of the majority of neuromuscular disorders.

The second difficulty is much more important. There is probably no other topic, in muscle pathology, in which exists a so large discrepancy between its clinical importance and our patho-physiological knowledge.

The "Index table" of all the textbooks - including *Myology* [2] - shows that almost all the research work is devoted to the understanding of the weakness or the atrophic process of the muscle, and almost nothing about the muscular contractures. However, these contractures represent one of the main subject for physiotherapists or surgeons in muscular dystrophies.

Table 1. Differences between English and French meanings

Contractures	English	French
	Spasticity (cerebral palsy, etc)	Contractures
With m. activity	Spasms (tetanus, neuromyotonia, dystonia, hysteria, etc)	Contractures
	Rigidity (malignant hyperthermia, etc)	Contractures
With electric silence	Post-exercise cramps "cramping" "stiffness" (McArdle's disease, etc)	Contractures
Without m. activity	Contractures (muscular dystrophies)	Retractions

Contractures are indeed invariably present in almost all neuromuscular disorders, but there is a great variability in their development. It is classically considered that two main factors contribute to their development :

1 - the rupture of the normal balance between agonists and antagonists muscles: the contractures develop in the muscles whose antagonists are becoming weaker and weaker;

2 - the development of interstitial fibrosis.

Historically, it is noticeable that the first one who emphasized the rupture of the balance agonists/antagonists, and demonstrated the progress of "myosclerosis" was Duchenne [3]. In Duchenne dystrophy, these two factors are indeed present. It is well known that contractures are first detected in hip flexors, ilio-tibial band and heel cords. They progress relatively slowly as long as the child is able to walk, but when he is confined to the wheelchair, contractures become prominent in the feet, around the hips, knees, elbows and wrist flexors. This underlines the role of active stretching and mobilization of articles in preventing the development of muscle contractures, as well as the therapeutic role of manual passive stretching of the muscles, specially of calves and ilio-tibial bands. A few authors tried to quantify the evolution of contractures *versus* the development of weakness, and demonstrate the rapid increase of the contractures when the child is spending the major part of the day in a wheelchair [1].

If the first factor – rupture of the balance between agonists and antagonists muscles may account for the distribution of the contractures and the type of subsequent deformities; the second factor - the development of interstial fibrosis - may account for the degreee of severity of the contractures in various muscular dystrophies.

To consider only the five main dystrophies characterized by a necrotic-regenerative pattern of lesions, there is for instance a scale of severity from facio-scapulo-humeral dystrophy to congenital muscular dystrophies. This "scale" is roughly correlated to the increase of endo- and perimysial collagen - which unfortunately was never precisely quantified in the different groups of muscle.

Furthermore, it is unclear why a so large range of intensity of the contractures exists within each entity. An example is the congenital muscular dystrophies in which early contractures may be the presenting symptom in merosin-deficient children, and which may present, in merosin-positive children, as Rigid Spine Syndroms, or as the so peculiar Ullrich syndrome.

But this is also true within a genetically well-identified entity, such as limb girdle muscular dystrophy, with calpain 3 gene involvement. In our experience, amongst twenty metropolitan French patients two were remarkable by an early and marked development of contractures in sural triceps, posterior thigh muscles and biceps brachialis - so marked that they were initially considered as close to Emery-Dreifuss humero-peroneal dystrophy ("between phenotype"). Mutations in the calpain gene are now determined. A difference with other French metropolitan cases is that both cases presented with null mutations, while other cases most often presented with one or two missense mutations [4].

This leads to consider the dystrophies in which an early development of "segmental" contractures is one of the distinctive sign of the disease. This is the case in Bethlem's myopathy, with contracture of the finger flexors. This is also the hallmark of Emery-Dreifuss muscular dystrophies [5]. In the early description by Cestan and Lejonne [6] of two brothers reported in 1904 in the *Nouvelle Iconographie de la Salpêtrière*, nothing is said about any heart dysfunction, but the phenotype is clearly similar to that described 64 years later by Emery and Dreifuss [7]. Two French families with an autosomal dominant form of the disease where recently studied. In the first large family existed a large phenotypic variability. Some members exhibited both cardiac and skeletal muscle involvement (two were heart-grafted), but several other showed only cardiac changes. The biopsies did not show any emerin deficiency. In a second family, contractures of the upper and lower limbs were typical of Emery-Dreifuss dystrophy but heart dysfunctions were limited to ECG abnormalities. It is noticeable that in theses families biceps brachialis contractures developed very early, in absence of any weakness of the triceps brachialis. A RMI study showed modifications in the biceps, but not in the triceps. Furthermore, deltoid muscle biopsies did not show any significant increase of endomysial collagen.

In these cases, the mechanisms which were proposed for the development of contractures in muscular dystrophies do not seem to be satisfying: contractures and weakness develop in parallel. Other mechanisms should be considered at the origin of these contractures.

One of the mechanisms which may be considered is in line with the demonstration of an adjustement of the number of sarcomeres to the functional length of the muscle fiber. This was studied years ago by Tardieu and coworkers [8] in collaboration with G. Goldspink [9]. Experimentally, they first demonstrated that there was a rapid adaptation in adult muscle fibers of the number of sarcomeres to the imposed distance between the tendinous insertions of the muscle. Immobilized in a shortened position, there was a quick decrease of the number of sarcomeres, and hence a considerable decrease in extensibility. The reverse was observed when the muscle was immobilised in lengthened position. Then looking to the mechanism of contractures in cerebral palsies, they found a similar displacement of the tension-extension curves and a marked diminution of the number of sarcomeres. This was considered as the basic cause of the contractures, of their irreductibility by manual passive stretch of the muscle, and of the necessity of a surgical lengthening of the tendon. This procedure made the passive mobilization easier, but without any benefit for the muscle strength on the whole range of the articulation.

Such a mechanism has never been explored in human muscular pathology. It is indeed very difficult to study this phenomenon in patients, due to the necessity to explore the muscle fibers tendon to tendon. However, such a mechanism might be explored in animal models of muscular such as mdx or dy/dy mices.

Another possibility is now available for studying the muscle contractures in human patients. It is indeed now possible to measure the muscle stiffness in muscular dystrophies. Experimentally, it was demonstrated that the Series Elastic Composents (SEC), in Hill's model, can be characterized by quick-release experiments. The same technic can be applied to muscular dystrophy patients. A study performed in our research group [10] indicate that an increase in stiffness index was correlated to the grade testing in Duchenne muscular dystrophy. This increase in stiffness may be explainable in terms of structural changes of the "active" structures of the muscle – the muscle fibers – and of the "passive" structures (tendon, surrounding connective tissue) which constitutes the SEC. This may represent a very useful tool for the clinical investigation of muscle contractures and the follow-up of muscular dystrophies.

Conclusions

These remarks only evoked some clues for further research in the understanding of "passive" muscle contractures; almost everything remains to be done in this very difficult, and clinically so important, field of research.

References

1. Brooke MH (1976) A clinician's view of Neuromuscular Diseases. Williams and Wilkins
2. Layzer RB (1994) Muscle pain, cramps and fatigue. In: Engel AG, Franzini-Armstrong C (eds) Myology. 2nd edn, McGraw Hill, tome II, chap 67, pp 1754-1768
3. Duchenne GBA (1868) Recherches sur la Paralysie Musculaire pseudo-hypertrophique ou paralysie myosclérosique. Archives Générales de Médecine 11, 5-25, 179-209, 305-321, 421-443, 552-588
4. Fardeau M, Eymard B, Mignard Cl, Tome RMS, Richard I, Beckmann JS (1996) Chromosome 15-linked limb Girdle Muscular Dystrophy: Clinical phenotypes in Reunion Island and French metropolitan communities. Neuromusc Disord 6:447-453
5 Rowland LP, Fetell M, Olarte M, Hays A, Singh N, Wanat FE (1979) Emery-Dreifuss Muscular Dystrophy. Ann Neurol 5:111-117
6. Cestan R, Lejonne (1904) Une myopathie avec rétractions familiales. Nouvelle iconographie de la Salpêtrière 15:37-52
7. Emery AEH, Dreifuss FE (1966) Unusual type of benign x-linked muscular dystrophy. J Neurol Neurosurg Psychiatry, 29:338
8. Tardieu G, Tabary JC, Tardieu C, Lombard M (1971) Rétraction, hyperextensibilité et "faiblesse" de l'IMC: expressions apparemment opposées d'un même trouble musculaire. Conséquences thérapeutiques. Rev Chir Orth 57:305-316
9. Tabary JC, Tabary C, Tardieu C, Tardieu G, Golspink G (1972) Physiological and structural changes in the cat i soleus muscle due to immobilization at different lengths by plaster casts. J Physiol 224:231-244
10. Cornu Ch, Goubel F, Fardeau M (1998) Stiffness of knee extensors in Duchenne muscular dystrophy. Muscle & Nerve 21:1772-1774

Contractures in Emery-Dreifuss Syndrome

K. Arahata, S. Fujimoto, E. Uyama, T. Ishikawa, S. Kubo, Y. Tsuchiya, I. Nonaka

Emery-Dreifuss syndrome (EDS) is an inherited muscular disorder characterized by the clinical triad of: (1) early contractures of the elbows, Achilles tendons, and postcervical muscles; (2) slowly progressive muscle wasting and weakness with a humeroperoneal distribution early in the course of the disease, and later on weakness spreads to the proximal limb muscles; and (3) a cardiomyopathy usually presenting as an atrio-ventricular block ranging from sinus bradycardia, prolongation of the PR interval to complete heart block. First description of EDS will back to 1902 when Cestan and Lejonne found brother patients with "rétractions musculaires". In 1966, Emery examined a large Virginian family affected with an X-linked muscular dystrophy [1]. The disease observed was quite distinct from both Duchenne and Becker muscular dystrophies (DMD and BMD); the unusual elbow and spine contractures, the proximal arm and distal leg pattern of weakness and the essential cardiac features. Similar symptoms were described in other families until 1979 when Rowland suggested the term "Emery-Dreifuss muscular dystrophy (EDMD)" for this distinctive disease.

Inheritance

In most cases, this disorder is inherited as an-X-linked recessive trait (McKusick 310300) (X-EDMD). X-EDMD shows clinical symptoms by the second-third decade of life, but heterozygous females show no indication of skeletal muscle disorder. However arrhythmia and bradycardia are occasionally observed in female carrier [2-5]. Rare autosomal dominant form (McKusik 181350, AD-EDMD) has also been reported [6-10]. AD-EDMD differs little clinically from X-EDMD in most cases [10-12], although in AD-EDMD families, age at onset and degreee of severity are more variable. It was therefore agreed that the diagnostic criteria for the EDMD phenotype previously established for X-EDMD could be applied to AD-EDMD [13, 14].

The X-EDMD gene was assigned to distal Xq in 1986 by demonstration of linkage with the FVIII gene and the anonymous marker DXS15.

Subsequent studies with more families and new markers placed the EDMD gene distal to DXS15 and refined the linkage in a region of about 2 Mb between the markers DXS15 and the factor VIII gene [15]. Recently, the autosomal dominant form has been mapped to chromosome 1g11-23 [16].

Recent advances in EDS

Identification of the *STA* gene, responsible for X-EDMD [17] and subsequent immunochemical analysis of the gene product [18, 19] have provided new insight into this life-threatening muscular disorder. The cDNA of STA contains an open reading frame of 762 nucleotides that encodes a novel 254 amino acid protein, named *"emerin"*, with a hydrophobic region of twenty amino acids near the C-terminus. Northern blot analysis shows that emerin is ubiquitously expressed, although stronger expression is observed in the skeletal muscle and heart. A single 1.2kb mRNA is expressed, and there has been no evidence of splicing variation. The human STA gene spans 2.1 kb and contains six exons and five introns. The 5' upstream (- 1 kb) region of the gene is rich in potential recognition sites for ubiquitous transcription factors, and CpG islands are also present. From its ubiquitous expression and promoter composition, *STA* is predicted to be a housekeeping gene, required particularly in tissues such as skeletal and cardiac muscles.

In AD-EDMD, mutations in the gene encoding lamin A/C have been identified [16].

Nuclear localization of emerin

Emerin is a novel protein showing a limited sequence similarity to the lamina associated protein 2 (LAP2), also known as thymopoietin beta [20, 21]. Both the N- and C-terminal region of emerin show similarity to LAP2 [22]. The C-terminal region contains a hydrophobic stretch of 30 amino acids that has the characteristics of a transmembrane domain. We raised two antisera (ED1 and ED2) against synthetic peptide fragments predicted from the emerin sequence, and have found positive immunostaining of the nuclear membrane in skeletal, cardiac and smooth muscle cells in normal controls and samples from a wide range of neuromuscular diseases such as Parkinson's disease, motor neuron diseases, metabolic myopathies, inflammatory myopathies, DMD and BMD [18]. On the other hand, no detectable emerin was observed in EDMD patients carrying premature stops in the ORF. Essentially the same results were found by Manilal et al. [19] who prepared a panel of monoclonal antibodies against emerin, and by Mora et al. [23] who used polyclonal antibodies raised against a truncated fusion protein.

Thus, the negative immunostaining for emerin in muscle fiber nuclear membrane in patients with X-EDMD was quite distinct from other neuromuscular diseases [24-26]. By Western blot. Emerin was identified as a 34 kd protein [17, 18, 21] slightly larger than the size estimated from the nucleotide sequence (29 kd). It is not clear whether this is due to post transcriptional modifications or to an aberrant migration during electrophoresis, although several consensus phosphorylation sites have been found. Importantly, lamin A/C also localizing at the nuclear membrane [16, 22].

Joint contractures

In patients with EDS, joint contractures are one of the major clinical hallmarks of the disease (Fig. 1).
The nature of the early onset joint contractures in the elbows, Achilles tendons, and posterior neck are not well characterized. Whether the contractures occur as a result of a primary abnormality of tissues that surround the joints, or are secondary to primary dystrophic changes in skeletal muscle, remains to be clarified. Merlini pointed out the early role of the primary muscular atrophy that may precede the contracture, as well as the value of CT scanning to confirm the pattern of selective muscle involvement particularly of the biceps, cervical and lumbar paravertebral muscles, and posterior thigh and leg muscles [14]. This clinical observation is very important and worth considering. However, in EDS, contractures often occur prior to significant weakness of muscle. Although EDS is quite distinctive from other forms of muscular dystrophy, several diseases require differential diagnosis from EDS. This includes rigid spine syndrome [26], congenital muscular dystrophies with joint contractures, scapulo-peroneal syndrome, Bethlem myopathy and other forms of X-linked muscular dystrophies. Recently, Cartegni et al. [27] found positive emerin staining at the cardiac intercalated discs, an adhesive junction, in control rat sample, and suggested a possible physiological role for the molecule as a cellular adhesion and/or communication between adjacent cells among the cardiac conduction system, although these findings have to be confirmed [28]. Many questions remain to be answered, however, and we do not know whether emerin interact with transmembrane adhesive glycoproteins that interact with extracellular matrix, thereby could produce joint contractures in EDS.

Conclusion and future perspective

Recent research on X-EDMD has given valuable insights into this life-treatening disorder, and raises several fundamental questions. Why does

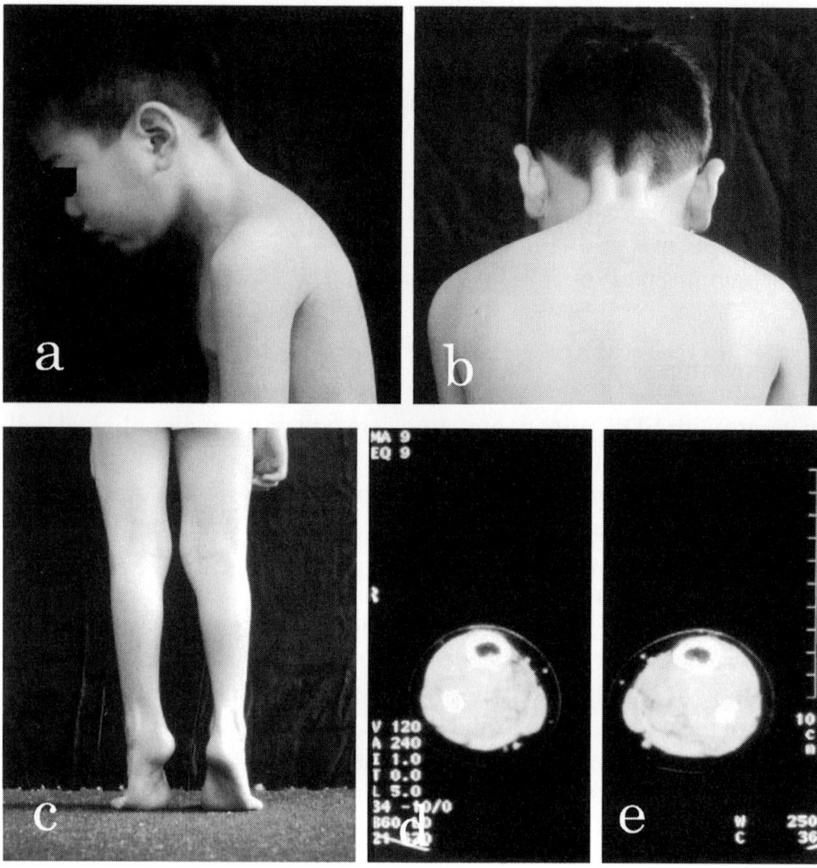

Fig. 1. A 7-year-old boy with emerin-deficient X-EDMD. He has marked limitation of his neck flexion (a) due to severe contractures in the posterior neck muscles (b). He shows toe walking and can not place his heels on the floor due to severe contractures in the Achilles tendons (c). CT scan shows increased low density in the both calves (d, e) and back muscle (not shown) with dystrophic changes in muscle (not shown)

the loss of an ubiquitous protein selectively affect skeletal muscles, heart and joints? Why does the dysfunction of a nuclear protein result in unique joint contractures? Is emerin functionally related to putative genes (or their products) for autosomal dominant type EDMD? Emerin is an element of a hitherto unknown nuclear membrane protein complex. On the basis of this assumption, the molecules interacting with emerin may also be important. It has been hypothesized that the putative product(s) of the gene responsible for AD-EDMD is one such molecule. Indeed, one of the molecules responsible for AD-EDMD has been identified as an inner nuclear membrane protein, lamin A/C. Further identification of homologous molecules will be informative.

Acknowledgements

We thank Drs. Hideo Sugita (Homorary President, NCNP, Japan) and Daniela Toniolo (Prof., CNR-IGBE, Italy) for their comments. This study has been supported in part by a Research Grant and - in - Aid for Brain Sciences from the Ministry of Health and Welfare, Japan.

References

1. Emery AEH, Dreifuss FE (1966) Unusual type of benign X-linked muscular dystrophy. J Neurol Neurosurg Psychiat 29:338-342
2. Merlini L, Granata C, Dominici P, Bonfi'glioli S (1986) Emery-Dreifuss muscular dystrophy: Report of five cases in a family and review of the literature. Muscle & Nerve 9:481-485
3. Pinelli G, Dominici P, Merlini L, Di Pasquale G, Granata G, Bonfiglioli S (1987) Cardiologic evaluation in a family with Emery-Dreifuss muscular dystrophy. G Itl Cardilo 17:589-593
4. Bialer MG, McDaniel NG, Kelly TE (1991) Progression of cardiac disease in Emery-Dreifuss muscular dystrophy. Clin Cardilo 14:411-416
5. Fishbein MC, Siegel RJ, Thompson CE, Hopkins LC (1993) Sudden death of a carrier of X-linked Emery-Dreifuss muscular dystrophy. Ann Intern Med 119: 900-905
6. Hauptman A, Thannhauser SJ (1941) Muscular shortening and dystrophy - a heterofamiliar disease. Arch Neurol Psychiat 46:654-664
7. Chakrabarti A, Pearce JMS (1981) Scapuloperoneal syndrome with cardiomyopathy: report of a family with autosomal dominant inheritance and unusual features. J Neurol Neurosurg Psychiatry 44:1146-52
8. Fenichel GM, Yi C-S, Kilroy AW, Blouin R (1982) An autosomal dominant dystrophy with humeropelvic distribution and cardiomyopathy. Neurology 32:1399-1401
9. Miller RG, Layzer RB, Mellenthin MA, Golabi M, Francoz RA, Mall JC (1985) Emery-Dreifuss muscular dystrophy with autosomal dominant transmission. Neurology 35:12301233
10. Witt TN, Graner CG, Pongratz D, Baur X (1988) Autosomal dominant Emery-Dreifuss syndrome: evidence of a neuroqMic variant of the disease. Eur Arch Psychiatry Neurol Sci 273:230-236
11. Emery AEH (1989) Emery-Dreifuss muscular dystrophy and other related disorders. British Med Bulle 45:772-787
12. Emery AEH (1993) Emery-Dreifuss syndrome. J Med Genet 26:637-641
13. Yates JRW (1991) European workshop on Emery-Dreifuss muscular dystrophy 1991. Neuromusc Disord 1:393-396
14. Yates JRW (1997) 43rd ENMC International Workshop on Emery-Dreifuss muscular dystrophy. Neuromusc Disord 7:67-69
15. Kress W, Muller E, Kausch K, Kullmann F, Mostacciuolo ML, Rietschel M, Rotthauwe HW, Schmalenberger B, Siciliano G, Voit T, Langenbeck U, Muller B, Danieli GA, Zerres K, Muller CR, Grimm T (1992) Multipoint linkage mapping of the Emery-Dreifuss muscular dystrophy gene. Neuromusc Disord 2:111-115
16. Bione G, Di Barletta MR, Varnous S, Becane H-M, Hammouda E-H, Merlini L, Muntoni F, Greenberg CR, Gary F, Urtizbera J-A, Duboc D, Fardeau M, Toniolo D, Schwartz K (1999) Mutations in the gene encoding lamin A/C cause autosomal dominant Emery-Dreifuss muscular dystrophy. Nature Genet 21:285-288
17. Bione S, Maestrini E, Rivella S, Mancini M, Regis S, Romeo G, Toniolo D (1994) Identification of a novel X-finked gene responsible for Emery-Dreifuss muscular dy strophy. Nat Genet 8:323-327

18. Nagano A, Koga R, Ogawa M, Kurano Y, Kawada J, Okada R, Hayashi YK, Tsukahara T, Arahata K (1996) Emerin deficiency at the nuclear membrane in patients with Emery-Dreifuss muscular dystrophy. Nat Genet 12:254-259
19. Manilal S, Nguyen TM, Sewry CA, Morris GE (1996) The Emery-Dreifuss muscular dystrophy protein, emerin, is a nuclear membrane protein. Hum Mol Genet 5:801-808
20. Furukawa K, Pant* N, Aebi U, Gerace L (1995) Cloning of a cDNA for lami-na-associated polypeptide 2 (LAP2) and identification of regions that specify targeting to the nuclear envelope. EMBO J 14:1626-1636
21. Harris CA, Andryuk PJ, Cline S, Chan KH, Natayan A, Siekierka JJ, Goldstein G (1994) Three distinct human thymopoietins are derived from alternatively spliced mRNAs. Proc Natl Acad Sci USA 91:6283-6287
22. Funakoshi M, Tsuchiya Y, Arahata K (1999) Emerin and Cardiomyopathy in Emery-Dreifuss muscular dystrophy. Neuromusc Disord 9:108-114
23. Mora M, Cartegni L, Di Blasi C, Barresi R, Bione S, Raffaele di Barletta M, Morandi L, Merlini L, Nigro V, Politano L, Donati MA, Conelio F, Cobianchi F, Toniolo D (1997) X-linked Emery-Dreifuss muscular dystrophy can be diagnosed from skin biopsy or blood sample. Ann Neurol 42:249-253
24. Sguarzoni S, Sabatelli P, Ognibene A, Toniolo D, Cartegni L, Cobianchi F, Petrini S, Merlini L, Maraldi NM (1998) Immunocytochemical detection of emerin within the nuclear matrix. Neuromusc Disord 8:338-344
25. Manilal S, Sewry CA, Nguyen thi Man, Muntoni F, Morris G (1997) Diagnosis of X-linked Emery-Dreifuss muscular dystrophy by protein analysis of leucocytes and skin with monoclonal antibodies. Neuromusc Disord 7:63-66
26. Fidziańska A, Toniolo D, Hausmanowa-Petrusewicz I (1998): Ultrastructural abnormality of sarcolemmal nuclei in Emery-Dreyfuss muscular dystrophy (EDMD). J Neurol Sci 159:88-93
27. Kubo S, Tsukahara T, Takemitsu M, Kim BY, Utsumi H, Nonaka I, Arahata K (1998) Presence of emerinopathy in cases of rigid spine syndrome. Neuromusc Disord 8:502-507
28. Cartegni L, Raffaele di Barletta M, Barresi R, Squarzoni S, Sabatelli P, Maraldi N, Mora M, Di Blasi C, Comelio F, Merlini L, Villa A, Cobianchi F, Toniolo D (1997) Heart-specific localization of emerin: new insights into Emery-Dreifuss muscular dystrophy. Hum Mol Genet 6:2257-2264

Arthrogryposis and Spinal Muscular Atrophy

T.L. Munsat

Both arthrogryposis multiplex congenita (AMC) and spinal muscular atrophy (SMA) are diseases which have been well characterized in both the past and recent literature. Although they began with quite disconnected histories, in recent years their paths have crossed in an interesting manner; the relationship newly defined by the tools of molecular genetics. I have been given the task of defining this history and relationship and have found it an interesting challenge. My contribution will be in the form of a critical review and new data will not be presented.

Arthrogryposis multiplex congenita

AMC can be defined as multiple joint fixation present at birth with normal, or reasonably normal, articular elements. There is firm consensus on this definition and agreement that AMC is not a single disease but a syndrome resulting from many different causes. In about 50% of patients with AMC the cause is a denervating/neurogenic process while the rest are associated with primary myopathies or spinal cord disease-usually myelomeningocoele. A small, but interesting number of patients have been demonstrated to have anti-achR antibodies [1].

There are several features which characterize AMC regardless of etiology. These include micrognathia, hip dislocation and reduced fetal intrauterine movement. The grypotic limbs may be fixed in a distorted or relatively normal position at birth. The more severe forms are incompatible with even short survival, while prolonged life is seen with the less severe condition. In patients who do survive, it is clear that the underlying neuromuscular condition is not progressive. Relapse after successful therapy is common. The neurogenic form of AMC is said to respond less well than other forms. Early surgical management of spinal deformity is important. Patients are usually not compromised intellectually.

The earliest description of AMC has been attributed to Otto [2] in a paper entitled *Monstrum Humanum Trunco Nimis Brevi et Extremitatibus Incurvatis*, a title which conveys the essence of the condition, even to those who are not Latin scholars. In 1899, Phocas severed the muscular tendons in a patient with AMC and noted that the joints remained fixed. The first report of histologic myopathic changes was in 1908 [3], while reductions in the number of anterior horn cells in AMC were documented in 1932 [4] and again in 1947 [5]. In the early 1940's, Hamburger [6] carried out a number of very innovative and interesting experiments by which grafted limbs were studied in culture. Among other observations he noted that disconnection of a limb from the CNS resulted in joint fixation. In an extensive 1951 review, James [7] noted that although the literature contained 240 clinically diagnosed cases of AMC, only 6 had had autopsies and of these only 3 had CNS examination. Nonetheless, by 1950 it had been reasonably agreed that AMC was the result of a number of different pathologic processes including neuropathic, myopathic and myelopathic conditions.

The hallmark studies in our modern understanding of AMC were initiated by Dan Drachman in the early 1960's [8]. The first paper in this series of incisive clinical and experimental work, was published in 1961 with Betty Banker [8]. After a detailed, analytic review of the literature, these authors reported a single classical case of AMC which had careful and exacting pathologic analysis.

The patient was a male born to a family without other evidence of neuromuscular disease. The mother had noted reduced fetal movements. At birth, extensive AMC was present, respiratory insufficiency developed and he died at 9 days of life. On histologic examination the muscles demonstrated classical denervation atrophy. Fixation of the joints was caused not by articular abnormalities but by periarticular soft tissue deposition. The spinal cord was noted to be narrow especially at the cervical and lumbar enlargements. The ventral roots were thinned. At the microscopic level there was a marked diminution in the number of large motor neurons in the anterior horn. Neuronophagia was only sparingly present, leading the authors to conclude that the pathologic damage had occurred early in fetal life and that an on-going degenerative process was not occurring. It was their view that most likely this damage occurred prior to 3-4 months of fetal development. Finally, of interest in the context of this particular discussion, was their analysis of a possible relationship between AMC and SMA. Although they noted that there were pathologic similarities between SMA and neurogenic AMC, they were more impressed by the clinical differences. For example they remarked that while AMC was a static, non-progressive condition, SMA was progressive. Ironically, it has now been well-documented that SMA is, in fact, non-progressive. In SMA surviors strength does not decrease over time and there is no evidence that additional motor units are lost as the SMA patient ages. Drachman and Banker suggested that "*In spite of these differences, we are unable to rule out the possibility that Werdnig-Hoffmann disease and the neural form of arthrogryposis may be related*" [8].

Following this report, Drachman [9] became intensely interested in the pathogenesis of AMC and devised an innovative animal model to test his hypotheses. Since respiratory gas exchange in the chick embryo occurs by passive diffusion across the chorioallantoic membrane, the neuromuscular junction could be paralyzed at an early stage without compromising further embryonic development. Specifically, rather substantial amounts of botulinum toxin, as well as other neuromuscular blocking agents, were injected during early development and the animal examined at time intervals thereafter. With botulinum induced neuromuscular paralysis an appearance very similar to human AMC was produced. The muscle showed changes typical of denervation atrophy while there was severe ankylosis of multiple joints. These experiments left no doubt in the minds of the authors that "paralysis per se rather than some incidental feature of the experimental procedure was responsible for the joint malformations". Paralysis as short as 24-48 hours was sufficient to cause the joint changes. The overall conclusion from these studies was that AMC resulted from early fetal limb inactivity per se and that any factor leading to paralysis of a limb was capable of inducing AMC. Drachman also suggested that the action of acetylcholine released from pre-synaptic terminals, as well as other as yet undefined humoral factors, resulted in a trophic effect on muscle which was essential to its functional survival. At this time there was growing support for the concept that many primary myopathies were, in fact, not caused by intrinsic defects in the muscle but by a lack of neuronal trophic factor(s). Drachman's animal studies and human observations firmly established the concept, which still holds to this day, that AMC is a syndrome of multiple causes which have in common limb paralysis early in fetal development.

Spinal muscular atrophy

In 1985 an interdisciplinary group of molecular biologists and clinicians set out to define the SMA gene. The initial effort was to establish a clinical definition of the disease that would be useful in defining a phenotypically homogeneous group of patients amenable to linkage analysis studies. After several meetings, sponsored primarily by ENMC, a consensus was achieved [10]. SMA was defined as follows:

Inclusions	
I - Weakness	II - Denervation
Symmetrical	By EMG
Proximal > distal	By muscle biopsy
Legs > arms	Clinical fasciculations
Trunk involved	
Exclusions	
CNS dysfunction	Sensory loss
Arthrogryposis	Eye muscle weakness
Involvement of other organs	Marked facial weakness

Childhood SMA, the main focus of interest, was divided into three catagories SMA I (severe) - onset prior to 6 months and death typically by 2 years; SMA II (intermediate) - onset by 18 months; SMA III (mild) - onset after 18 months.

At the time these criteria were developed it was still unclear as to whether SMA was a single disease with phenotypic variability or several different genetic entities. Although arthrogryposis was listed as an exclusion, there were some involved in the discussions who believed that severe SMA could present at birth with AMC. However, for the sake of assuring a homogeneous group of patients for linkage studies neurogenic AMC was excluded.

Genetic linkage to chromosome 5q11.2-13.3 was established in 1990 by a group at Columbia University led by Conrad Gilliam [11]. Almost immediately additional linkage studies began to sharpen the phenotypic definition of SMA. Justification of the decision to exclude patients with significant CNS damage was reported by Dubowitz [12] in a patient with "SMA" and olivopontocerebellar hypoplasia. Similarly, patients with hexosamindase deficiency and "SMA" were shown to have a different disease [13]. A large consanguinous family with SMA and AMC did show linkage to the SMA locus [14].

In 1995 Judith Melki's lab in Paris [15] defined a gene in the suspect region which was partially or completely deleted in over 95% of Type I and Type II SMA patients which she named "survival motor neuron" gene - SMN. Other groups have now confirmed that defects in the SMN gene, whose function has not yet been completely determined, is the main determinant of the disease, although other genes may play a contributing role in some patients. The discovery of the gene responsible for SMA now allowed a more precise definition of the disease. For example, although cardiac defects were initially considered an exclusion for the diagnosis, it may be a manifestation of SMA in some patients [16]. Neurogenic AMC can also be seen on rare occasions with an x-linked pattern of inheritance. These patients present with hypotonia, areflexia and arthrogryposis at birth associated with loss of anterior horn cells and early death. This disease maps to the Xp11.3-q11.2 region [Kobayashi, 1995].

In 1996 [17] the Melki team reported that the SMN gene was defective in only 50% (6 of 12) of neurogenic AMC patients. This firmly established the fact that SMA could indeed present with AMC. But possibly of greater interest these findings demonstrated that there were other neurogenic conditions, phenotypically identical to SMA, that could also present with AMC. The nature of these conditions is unknown. The following year Bingham [18] using DNA extracted from paraffin blocks, reported that exons 7 and 8 of the SMN gene were deleted in two of four infants with neurogenic AMC, confirming the Melki report that 50% of neurogenic AMC patients had a disease other than SMA.

Conclusion

AMC is a syndrome of multiple causes. Lack of limb movement early in fetal development is the common element in its pathogenesis. SMA can present with AMC, but 50% of neurogenic AMC remains unexplained.

References

1. Hall JG (1997) Arthrogryposis multiplex congenita: etiology, genetics, classification, diagnostic approach, and general aspects. J Pediatr Orthop B 6:159-66
2. Otto, cited by James T (1951) Multiple congenital articular rigidities. Edinburgh Med J 58:565-597
3. Howard RA (1908) case of congenital defect of the muscular system (dystrophia muscularis congenita) and its association with congenital talipes equinovarus. Proc Roy Soc Med 2:157-166
4. Price DS (1932)A case of amyoplasia congenita, with pathological report. Arch Dis Child 8:343-354
5. Brandt S (1947) Arthrogryposis multiplex congenita. Acta Pediat 34:365-381
6. Hamburger V, Waugh M (1940) The primary development of the skeleton in nerveless and poorly innervated limb transplants of chick embryos. Physiol Zool 13:367-380
7. James T (1950) Multiple congenital articular rigidities. Edinburg Med J 58:565
8. Drachman DB, Banker BQ (1961) Arthrogryposis multiplex congenita. Case due to disease of the anterior horn cells. Arch Neurol 5:77-93
9. Drachman DB (1969) Congenital deformities produced by neuromuscular disorders of the developing embryo. In: Motor Neuron Diseases Grune and Stratton, Inc
10. Munsat TL, Davies KE (1992) International SMA consortium meeting. 26-28 June, Bonn, Germany. Neuromusc Disord 2:423-428
11. Brzustowicz LM, Lehner T, Castilla IH et al (1990) Genetic mapping of chronic childhood-onset spinal muscular atrophy to chromosome 5q11.2-13.3. Nature 344:540-541
12. Dubowitz V, Daniels RJ, Davies KE (1995) Olivopontocerebellar hypoplasia with anterior horn cell involvement (SMA) does not localize to chromosome 5q. Neuromusc Disord 5:25-29

13. Kleyn PW, Brzustowicz LM, Wilhelmsen KC et al (1991) Spinal muscular atrophy is not the result of mutations at the beta-hexosaminidase or GM2 activator loci. Neurology 41:1418-1422
14. Shohat M, Lotan R, Magal N et al (1997)A gene for arthrogryposis multiplex congenita neuropathic type is linked to D5S394 on chromosome 5qter. Am J Hum. Genet 61:1139-1143
15. Lefebvre S, Burglen L, Rebullet S et al (1995) Identification and characterization of a spinal muscular atrophy-determining gene. Cell 80:155-165
16. Burglen L, Spiegel R, Ignatius J et al (1995) SMN gene deletion in variant of infantile spinal muscular atrophy. Lancet 346:316-317
17. Burglen L, Amiel J, Viollet L et al (1996) Survival motor neuron gene deletion in the arthrogryposis multiplex congenita-spinal muscular atrophy association. J Clin Invest 98:1130-1132
18. Bingham PM, Shen N, Rennert H et al (1997) Arthrogryposis due to infantile neuronal degeneration associated with deletion of the SMN gene. Neurology 49:848-851

Therapy of Muscular Contractures

B. Chabrol, P. Carrelet, G. Bollini

Contractures occur early in the course of neuromuscular diseases. They may even be present at birth as in the case of congenital myotonic dystrophy and congenital myopathies. Up till now, contractures were regarded as an exclusion criterion in spinal muscular atrophy (type I SMA), but recently a deletion of the survival motor neuron gene was found in some cases of type I SMA [1, 2]. When multiple joint contractures are present at birth, the term Arthrogryposis is used [3].

In progressive muscular dystrophy contractures appear at different stages in the course of the disease. Contractures on weight bearing joints seem to be the most important factor, besides muscular weakness causing loss of walking. Contractures must be systematically traced and treated. In fact, contractures have a pernicious influence not only on functional abilities but also on home care and pain.

Development of contractures

Contractures are disabling complications throughout the course of neuromuscular disease. They occur due to muscular weakness and as a result of imbalance between agonists and antagonists in both neurologic diseases and muscular diseases.

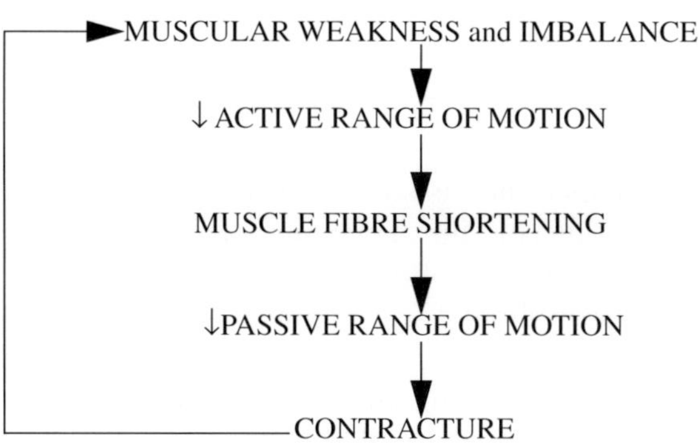

A vicious circle appears to be initiated by asymmetrical muscular weakness and imbalance resulting in first a decrease of active range of motion, then a shortening of the muscle fibre, then restriction of passive range of motion, and finally contracture. So, contractures cause a decrease in the active range of motion. At the same time, as weakness is not quite symmetrical, imbalance on frontal level induces asymmetrical standing or sitting and aggravates scoliosis.

Imbalance usually develops towards flexors that benefit while abdominal muscles and extensors get weaker. Hips become fixed in an abduction, flexion, external turning pattern; knees get fixed in a flexor pattern and ankles in equinus because of hamstrings and gastrocnecmius prevalence.

Contractures according to disease

Congenital myotonic dystrophy

Contractures are usually present at birth, setting up at the most arthrogryposis pattern with: club foot, ulnar wrist imbalance, elbow flexor, hip flexor or extensor, knee flexor or extensor. It gets vital to immediately manage these contractures because, if managed soon, they are dramatically resolving and walking may be acquired later on by these children.

Congenital myopathy

In nemalin myopathy, contractures are also present at birth and concern foot, hands and hips. In centronuclear myopathy club foot frequently occur [4]. In congenital myopathy, patterns of fetal hypokinesia or arthrogryposis are described.

Congenital muscular dystrophy

Early histologic conjunctive infiltration conducts to early contractures and leads to great deformities and sometimes even to arthrogryposis because of fetal immobility. Elsewhere extreme weakness leads to hip luxation which is often recurrent, and to equinus and cyphoscoliosis [4, 5].

Spinal muscular atrophy

Type I induces deformities either as soon as birth or quickly advancing like wrist imbalance, equinus or talus, knee flexor and valgum, hip flexor with oblique tilt damaging early scoliosis.

Type II is often diagnosed about 9 or 12 months. At this time, knee flexor/ valgum and equinus or more rarely. Talus/valgus are observed.

Type III allows usually ambulating until 15 years or even more. However, walking is possible at the expense of "functional" deformities. They can be reduced when lying and must be tolerated.

Other diseases

In neonatal myasthenia or in congenital myasthenia, arthrogryposis can be observed at birth [6].

Hereditary motor and sensitive neuropathies

Contractures due to imbalance between intrinsic muscles, which are weak, and long muscles that are comparitively spared, leads to toe flexor and cavus feet. It gets quickly stiff and often requests early surgery. On the level of the hands, same imbalance between short and long muscles leads to finger flexor.

Progressive muscular dystrophy

The child's joint often look like normal until the age of 6. At most, equinus due to calf contracture is already present. Later on, hip flexor appears because of gluteus and abdominal muscle weakness. Contractures stay rather soft as long as the child keeps on walking or standing. When walking is over, contractures get stiff and in first affect the weight bearing joints. Upper limb is rather spared during most course of disease.

Goals of therapy

Goals should be to give the patient a life as normal as possible, both physically and socially for as long as possible. This means that the child would be able to acquire walking or to try to live an autonomous life as much as possible. It requests a good alignment of weight bearing joints, with or without orthoses. It also supposes that muscular strength should be preserved at its best amount [7].

It requests an early diagnosis in a quite definitive way. As already stated, contractures in congenital myotonic dystrophy and in some congenital myopathy may get reduced or resolved if early treated. On the other hand, contractures progressively appear and get worse in a progressive muscular disease such as Duchenne muscular dystrophy.

Management of patient

Patient's management with neuromuscular diseases must be carried out by a multidisciplinary team in order to define the best care. In fact, a combination of a severe locomotion disturbance, psychological trouble and sometimes mental retardation requests on help of neuropediatrician, orthopaedic surgeon, rehabilitation doctor, psychologist, physiotherapist and social workers [7]. Indeed, because of child's growth and for the progressive course of disease the social and orthopaedic status of the patient is always changing and very interdependent [8]. Also using a wheel-chair is often refused in some schools or in children's holiday camps for example. After loss of ambulating, the patient's life should stay as comfortable as possible. This supposes light technique aids, an electric wheel-chair and no joint or muscular pain. In this way, "triscooter" will be better received by teenagers than a standard electric wheel-chair. A triscooter needs a good trunk control. Some electric wheel-chairs allow standing and moving at the same time. In our opinion, early choice of a standing wheel chair at the very moment of loss of walking is better than by all means trying to keep a child walking, for, in either way, the ability of walking will be lost.

Means and methods

Muscular exercise

Some years ago, Vignos described the effects of exercise in muscular dystrophy [9, 10]. He proposed a rehabilitation program which consisted of muscular exercises such as hip abduction, hip extension, knee extension, forward flexion of the arm and sit-ups for abdominal strengthening. Weights were even used if muscular strength was sufficient. He concluded that for ambulatory patients, this program improves the amount of weight lifted, and, especially in facios-capulo-humeral dystrophy, improves functional activities such as stair climbing or gait stability. In the other way, other authors decided to forbid any muscular exercise in Duchenne muscular dystrophy in order to preserve residual muscular amount. They argue that for healthy and trained people muscular stress as an intensive and durable effort (marathon, for example) induces muscular fibres necrosis. For healthy young people regeneration begins early after effort and is complete. In muscular dystrophy, necrosed fibres are supposed not to regenerate. That is the reason why they forbid muscular exercise [11]. Thus, debate about benefit or damage of muscular exercise in neuromuscular disease is still actual [12]. It is however usually considered that muscular exercise, and even sportive exercise may be pleasant, safe and useful to prevent disuse atrophy. It should exclude endurance and pain. In that way, swimming, horseriding, cycling, sailing or chair foot should be available for the child. Sportive exercise contributes to a normal range of motion, prevents social exclusion, but over all gives enjoyment and pleasure to the child.

Physiotherapy

Physiotherapy is often the first therapy prescribed [13, 14]. Later, it will continue with splinting or orthopaedic surgery. It begins at the time of diagnosis and lasts as long as the disease. It requires the child's agreement in spite of its tedious feature. It confronts the child again with its disease and, to be efficient, requires a lot of time per day or per week at the beginning.

A physiotherapist should be a confident person, a friend as well as a psychologist, but in any way always a therapist.

Massages seem to be no longer up to date. However, a physiotherapist should always start a massage sitting. In this way, he wins the child's confidence and is able to soften stiff tissues. He can make stretching easier and less painful.

Stretching may be active or passive.

Active stretching must be used as long as it is possible to perform muscular strength. It is very useful to prevent contractures. It consists of heel walking to prevent equinus, active range of motion and especially hip extension, knee extension and elbow extension. Walking is also a way of active stretching.

Passive stretching is used to reinforce range of motion which is not complete during active stretching or when active exercise is no longer possible. Physiotherapists have to find a balance between efficiency and pain.

Positioning is a way of stretching. It may be standing positioning that prevents or delays hip, knee and ankle contractures. It may also consists of proning that prevents hip flexor.

Passive range of motion

A passive range of motion is early necessary to prevent or limit contractures. As seen before, muscular weakness no longer permit a complete active range of motion. Secondary, muscular length decreases carrying out loss of function and so on. A passive range of motion is vital in neonatal diseases since no active range of motion is possible for a newborn infant [15]. In progressive muscular dystrophy, as long as walking is preserved, active stretching and positioning are sufficient.

Balneotherapy

Even for children who do not swim, that is a pleasant way of physiotherapy. Water temperature must be about 32 - 35° C. Sitting must not exceed 30 minutes. A range of motion is often easier in water and less uncomfortable, especially with arthrogryposis features.

Splinting and orthoses

As soon as contractures appear, they have to be systematically tackled. Orthoses achieve active or passive positioning.

Active orthoses supply muscular weakness in order to preserve a function. Long orthoses such as Jarrige or C.D.O. orthoses preserve walking for several months in Duchenne muscular dystrophy [16]. Short orthoses like "calf-ankle" struggle equinus and stepping especially in Charcot-Marie neuropathy (Fig. 1a, 1b). Elsewhere, they prevent knee recurvator as in congenital myopathy. Standing is enabled with long orthoses that are usually fixed on a brace (Fig.2). They are used in spinal muscular atrophy, in congenital myopathy and congenital muscular dystrophy. In fact, in several of these diseases, walking is either never acquired or quickly lost so that standing is only possible with these orthoses. Hip and knee locks allow sitting and standing alternance, especially in a standing electric wheel-chair. Passive orthoses are essentially used during the night achieving positioning while sleeping. These days, many splinting materials still exist. Gyps have been replaced by thermoplastic splinting or more recently by resin splinting which are lighter and more comfortable [17]. They are prescribed to fight against: hip flexor, asymmetrical tilt of hips, equinus or talus, ulnar imbalance of wrist (Fig 3a, 3b). These splintings are used for all neuromuscular diseases. It could be used at birth, as well as in congenital myotonic dystrophy, or spinal muscular atrophy (type I). In neuropathy or in progressive muscular dystrophy they are used later on, when contractures appear. They are better received than used when contractures are not stiff.

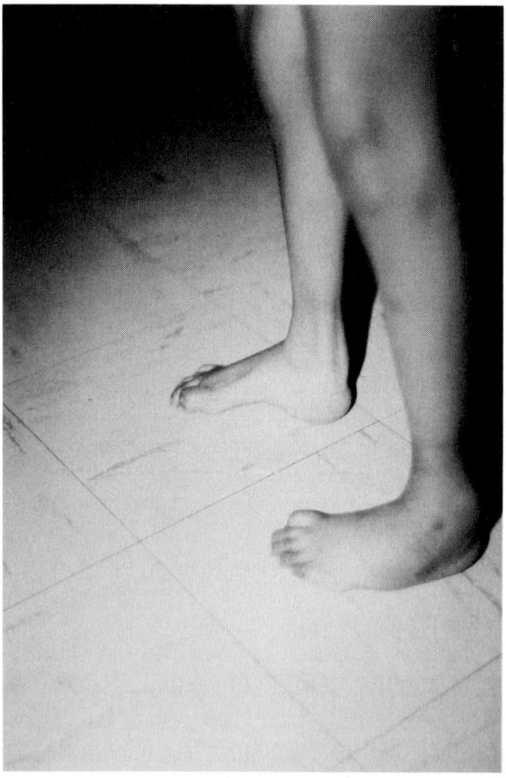

Fig. 1a. This young boy suffering Charcot-Marie neuropathy, presents cavus varus foot

However, because of contracture evolution, stretching, positioning, and splinting are often no longer sufficient to preserve a range of motion. As the child grows, contractures get worse and surgery is obviously necessary.

Fig. 1b. For this same young boy with Charcot-Marie neuropathy, orthoses prevent foot deformations

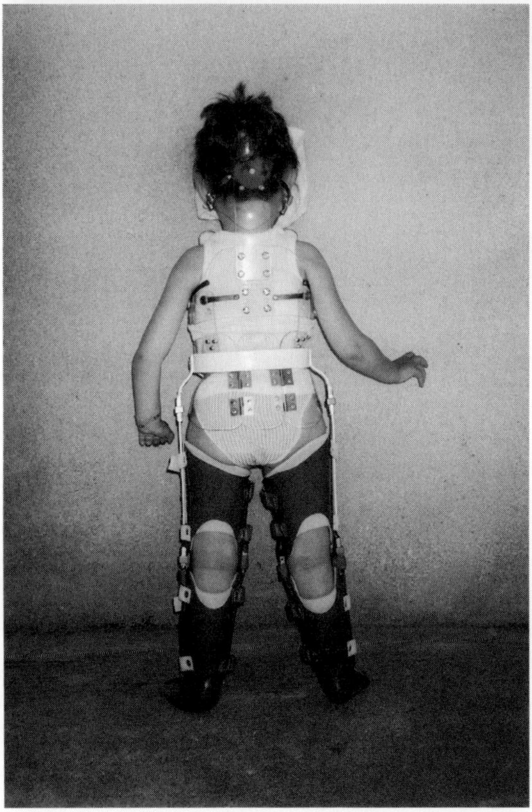

Fig. 2. For this young boy with spinal muscular atrophy type II, leg orthoses are fixed on the "garchois" brace, and allow verticalisation some hours per days. Locks on hip and knee allow sitting.

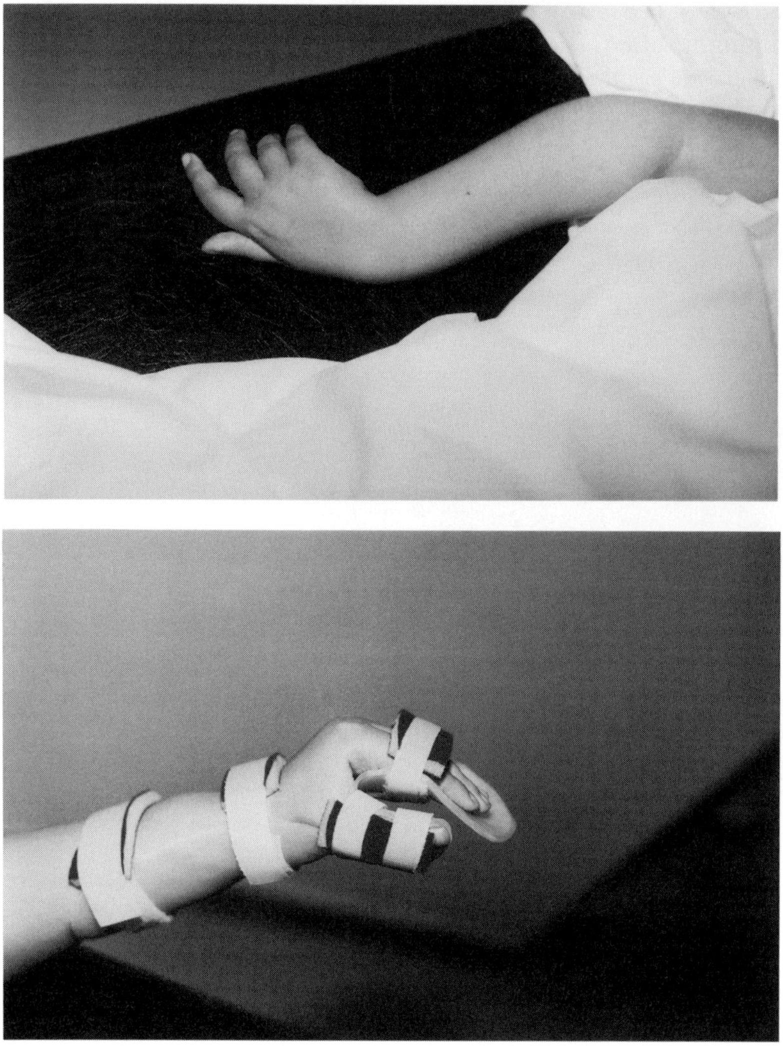

Fig 3. a) This young girl suffering spinal muscular atrophy type I presents flexor of wrist; b) for this same girl, orthoses prevent wrist contracture

Surgery

When contractures are fixed, only surgery may restore muscular, or rather tendon length. As seen before, normal muscular length is necessary to preserve function, harmonious growth, or at least a range of motion sufficient to dress, to put on shoes, to nurse and to keep a comfortable way of sitting. Most recent operations are subcutaneous tenotomies, especially at Achilles'level, hamstring, fascia lata tensor, hip flexors and toe flexors. These operations are performed alone or at the same time of spine arthrodesis.

In a preventive goal, some authors proposed early and wide surgery on hip and knee to preserve the ability of walking [18, 19]. These kinds of operation are efficient since a few months, so in comparison between the benefit and the heaviness of care, our team does not propose these operations.

As to conclude, muscular contractures are an available part of the natural history of neuromuscular disease. It occurs more or less soon according to the type of disease. It must be tracked in order to be managed as soon as possible. There is no single way of care. Patient's management must be discussed and defined, every case within a multidisciplinary team that gathers neuropediatrician, rehabilitation doctor, orthopaedic surgeon, physiotherapist, psychologist and social workers. The main goal is to give to these children a lot of autonomy, comfort and over all a life as pleasant as possible in spite of the disease.

References

1. Bingham PM, Shen N, Rennert H, Rorke LB, Black AW, Padilla M, Nordgren RE (1997) Artrhogryposis due to infantile neuronal degeneration associated with deletion of the SMNT gene. Neurology 49:848-851
2. Burglen L, Amiel J, Viollet L et al (1996) Survival motor neuron gene deletions in the arthrogryposis multiplex congenital spinal muscular atrophy association. J Clin Invest 98:1130-1132
3. Hall JG (1997) Arthrogryposis multiplex congenita: etiology, genetics, classification, diagnostic approach and general aspects J Pediatr Orthop B 6:159-166
4. Routon MC (1998) Myopathies congénitales. Dystrophies musculaires congénitales. In: Arthuis M, Dulac O, Mancini J, Pinsard N, Ponsot G (eds) Neurologie Pédiatrique. Flammarion Medecine-Sciences, Paris, pp 802-816
5. Dubowitz (1978) Muscle disorder in childood. Major problems in clinical pediatrics. Vol XVI, Saunders
6. Vajsar J, Sloane A, Mac Gregor DL, Ronen DGM, Becker LE, Jay V (1995) Arthrogryposis multiplex congenita due to congenital myasthenic syndrome. Pediatr Neurol 12:237-241
7. Vignos PJ, Spencer G, Archibald KC (1963) Management of progressive muscular dystrophy of childood. JAMA 184:103-110
8. Pous JG (1985) Marche et déplacement de l'enfant paralysé. In: Dimeglio A (eds) "L'enfant paralysé: rééducation et appareillage" Problèmes en Médecine de Rééducation. Masson, vol 6, pp 149-155
9. Vignos PJ (1983) Physical models of rehabilitation in neuromuscular disease. Muscle and Nerve 6:323-338
10. Vignos PJ, Watkins MP (1966) The effect of exercise in muscular dystrophy. JAMA 197:121-126
11. Urtizberea JA, Gilardeau G (1996) Exercice musculaire et maladies neuromusculaires. In: Pelissier JF, Urtizberea JA (eds) Les maladies neuromusculaires: de la génétique à la réadaptation. Masson, pp 323-327
12. Iannacone ST, White M, Browne R, Russman B, Buncher R, Samaha FJ (1997) Muscle fatigue in spinal muscular atrophy. J Child Neurol 12:321-326

13. Ledoux P, Larroque A (1995) Amyotrophies spinales infantiles. Kinésithérapie de l'enfant paralysé. In: Kinesithérapie pédiatrique. Vol 4, Masson pp109-118
14. Ledoux P, Larroque A (1995) La maladie de Duchenne de Boulogne. Kinésithérapie de l'enfant paralysé. In: Kinesithérapie pédiatrique. Vol 4, Masson pp 119-144
15. Muray C, Fixsen JA (1997) Management of knee deformity in classical arthrogryposis multiplex congenita (amyoplasia congenita), J Pediatr Orthop B 6:186-191
16. Beltramo F, Fort D, Dauphin D (1996) Appareillage de marche du myopathe: Réalités et perspectives. In: Pelissier JF, Urtizberea JA (eds) Les maladies neuromusculaires: de la génétique à la réadaptation. Masson, pp 143-150
17. Carrelet P, Chabrol B, Jouve JL (1998) Association de résine semi-rigide et de résine rigide: application aux orthèses en rééducation et aux pieds bots de l'enfant. In: Dimeglio A (eds) Le pied de l'enfant et de l'adolescent. Coll de Pathologie Locomotrice et de Medecine Orthopedique. Masson, pp356-358
18. Forst R, Forst J (1995) Importance of lower limb surgery in Duchenne muscular dystrophy Arch OrthopTtrauma Surg 114:106-111
19. Rideau Y, Duport G, Delaubier A et al (1996) Correction précoce des inégalités musculaires dans la myopathie: analyse internationale. In: Pelissier JF, Urtizberea JA (eds) "Les maladies neuromusculaires: de la génétique à la réadaptation". Masson, pp 156-170

Subject Index

CET OUVRAGE A ÉTÉ ACHEVÉ D'IMPRIMER EN JUIN 1999
SUR LES PRESSES DE L'IMPRIMERIE DE L'INDÉPENDANT - 53200 CHÂTEAU-GONTIER
DÉPÔT LÉGAL : 2e TRIMESTRE 1999